Principles of Virology

Principles of Virology

Lincoln Davenport

STATES
ACADEMIC PRESS
www.statesacademicpress.com

Principles of Virology
Lincoln Davenport
ISBN: 978-1-63989-443-7 (Hardback)

STATES
ACADEMIC PRESS

Published by States Academic Press,
109 South 5th Street,
Brooklyn, NY 11249, USA

Cataloging-in-Publication Data

Principles of virology / Lincoln Davenport.
p. cm.
Includes bibliographical references and index.
ISBN 978-1-63989-443-7
1. Virology. 2. Viruses. 3. Microbiology. I. Davenport, Lincoln.
QR360 .P75 2022
579.2--dc23

This book has been written, keeping in view that students want more practical information. Thus, my aim has been to make it as comprehensive as possible for the readers. I would like to extend my thanks to my family and co-workers for their knowledge, support and encouragement all along.

A virus is an infectious agent that can multiply inside the living cells of an organism. The scientific discipline concerned with the study of virus is called virology. This discipline is focused on the structure, interactions, ecology, evolution, physiology and clinical aspects of viruses. Virus classification refers to the process of identifying and classifying viruses into a taxonomic system. They are majorly classified into animal virus, plant virus, etc. Virology also studies subviral particles like viroids and satellites, which are smaller and simpler than viruses. Viral diseases are a major focus area under virology. Viral disease or infection occurs when an organism's body is infected by pathogenic and infectious viruses and related particles. Such diseases are common cold, HIV/AIDS, parainfluenza virus infection, etc. This book studies, analyzes and upholds the pillars of virology and its utmost significance in modern times. Different approaches, evaluations, methodologies and studies in this field have been included in this book. The topics covered in this textbook offer the readers new insights in the field of virology.

A brief description of the chapters is provided below for further understanding:

Chapter - Introduction

Small infectious agent that duplicates only inside the living cells of an organism is termed as a virus. The scientific discipline that is associated with the study of biology of viruses and viral diseases is called virology. This is an introductory chapter which will briefly introduce about virology.

Chapter - Branches of Virology

There are different branches of virology which can be classified as clinical virology, synthetic virology, neurovirology, veterinary virology, paleovirology and molecular virology. This chapter has been carefully written to provide an easy understanding of the various branches of virology.

Chapter - Classification of Viruses

Viruses can be categorized depending on the host cell they interact with. It is classified into plant viruses, animal viruses, fungal viruses and bacteriophages. The topics elaborated in this chapter will help in gaining a better perspective of the classification of viruses.

Chapter - Viral Life Cycle

The life cycle of viruses follow several stages for infecting the host cells. Its stages are attachment, penetration, uncoating, biosynthesis, maturation and release. Its other stages are viral entry, replication, shredding and latency. This chapter closely examines viral life cycle to provide an extensive understanding of the subject.

Chapter - Viral Proteins

Viral proteins are the products of viruses that are grouped as per their functions and structures. It also includes structural proteins, nonstructural proteins, regulatory, and accessory proteins. This chapter delves into the subject of viral proteins for an in-depth understanding of the subject.

Chapter - Viral Diseases

Viral diseases occur when an organism's body is attacked by pathogenic viruses, and virions. Some diseases are chicken pox, bronchiolitis, HIV/ AIDS, etc. All the aspects related to viral infections and diseases have been carefully analyzed in this chapter.

Lincoln Davenport

Introduction 1

Small infectious agent that duplicates only inside the living cells of an organism is termed as a virus. The scientific discipline that is associated with the study of biology of viruses and viral diseases is called virology. This is an introductory chapter which will briefly introduce about virology.

Virus

Virus is an infectious agent of small size and simple composition that can multiply only in living cells of animals, plants, or bacteria.

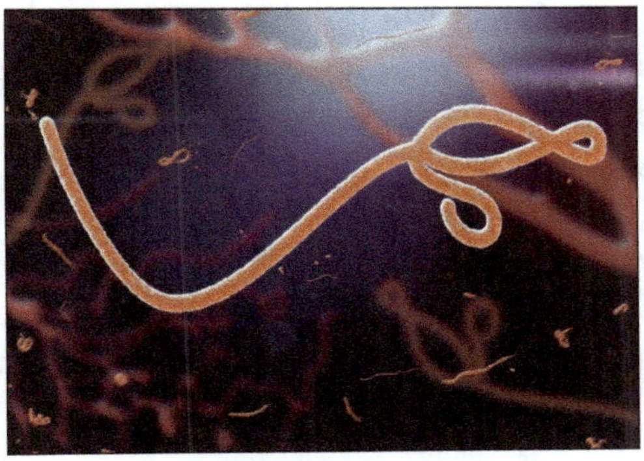

Ebola virus.

The earliest indications of the biological nature of viruses came from studies in 1892 by the Russian scientist Dmitry I. Ivanovsky and in 1898 by the Dutch scientist Martinus W. Beijerinck. Beijerinck first surmised that the virus under study was a new kind of infectious agent, which he designated contagium vivum fluidum, meaning that it was a live, reproducing organism that differed from other organisms. Both of these investigators found that a disease of tobacco plants could be transmitted by an agent, later called tobacco mosaic virus, passing through a minute filter that would not allow the passage of bacteria. This virus and those subsequently isolated would not grow on an artificial medium and were not visible under the light microscope. In independent studies in

1915 by the British investigator Frederick W. Twort and in 1917 by the French Canadian scientist Félix H. d'Hérelle, lesions in cultures of bacteria were discovered and attributed to an agent called bacteriophage ("eater of bacteria"), now known to be viruses that specifically infect bacteria.

The unique nature of these agents meant that new methods and alternative models had to be developed to study and classify them. The study of viruses confined exclusively or largely to humans, however, posed the formidable problem of finding a susceptible animal host. In 1933 the British investigators Wilson Smith, Christopher H. Andrewes, and Patrick P. Laidlaw were able to transmit influenza to ferrets, and the influenza virus was subsequently adapted to mice. In 1941 the American scientist George K. Hirst found that influenza virus grown in tissues of the chicken embryo could be detected by its capacity to agglutinate (draw together) red blood cells.

A significant advance was made by the American scientists John Enders, Thomas Weller, and Frederick Robbins, who in 1949 developed the technique of culturing cells on glass surfaces; cells could then be infected with the viruses that cause polio (poliovirus) and other diseases. (Until this time, the poliovirus could be grown only in the brains of chimpanzees or the spinal cords of monkeys.) Culturing cells on glass surfaces opened the way for diseases caused by viruses to be identified by their effects on cells (cytopathogenic effect) and by the presence of antibodies to them in the blood. Cell culture then led to the development and production of vaccines (preparations used to elicit immunity against a disease) such as the poliovirus vaccine.

Scientists were soon able to detect the number of bacterial viruses in a culture vessel by measuring their ability to break apart (lyse) adjoining bacteria in an area of bacteria (lawn) overlaid with an inert gelatinous substance called agar—viral action that resulted in a clearing, or "plaque." The American scientist Renato Dulbecco in 1952 applied this technique to measuring the number of animal viruses that could produce plaques in layers of adjoining animal cells overlaid with agar. In the 1940s the development of the electron microscope permitted individual virus particles to be seen for the first time, leading to the classification of viruses and giving insight into their structure.

Advancements that have been made in chemistry, physics, and molecular biology since the 1960s have revolutionized the study of viruses. For example, electrophoresis on gel substrates gave a deeper understanding of the protein and nucleic acid composition of viruses. More-sophisticated immunologic procedures, including the use of monoclonal antibodies directed to specific antigenic sites on proteins, gave a better insight into the structure and function of viral proteins. The progress made in the physics of crystals that could be studied by X-ray diffraction provided the high resolution required to discover the basic structure of minute viruses. Applications of new knowledge about cell biology and biochemistry helped to determine how viruses use their host cells for synthesizing viral nucleic acids and proteins.

The revolution that took place in the field of molecular biology allowed the genetic information encoded in nucleic acids of viruses—which enables viruses to reproduce, synthesize unique proteins, and alter cellular functions—to be studied. In fact, the chemical and physical simplicity of viruses has made them an incisive experimental tool for probing the molecular events involved in certain life processes. Their potential ecological significance was realized in the early 21st century, following the discovery of giant viruses in aquatic environments in different parts of the world.

General Features

Viruses occupy a special taxonomic position: they are not plants, animals, or prokaryotic bacteria (single-cell organisms without defined nuclei), and they are generally placed in their own kingdom. In fact, viruses should not even be considered organisms, in the strictest sense, because they are not free-living; i.e., they cannot reproduce and carry on metabolic processes without a host cell.

All true viruses contain nucleic acid—either DNA (deoxyribonucleic acid) or RNA (ribonucleic acid)—and protein. The nucleic acid encodes the genetic information unique for each virus. The infective, extracellular (outside the cell) form of a virus is called the virion. It contains at least one unique protein synthesized by specific genes in the nucleic acid of that virus. In virtually all viruses, at least one of these proteins forms a shell (called a capsid) around the nucleic acid. Certain viruses also have other proteins internal to the capsid; some of these proteins act as enzymes, often during the synthesis of viral nucleic acids. Viroids (meaning "viruslike") are disease-causing organisms that contain only nucleic acid and have no structural proteins. Other viruslike particles called prions are composed primarily of a protein tightly complexed with a small nucleic acid molecule. Prions are very resistant to inactivation and appear to cause degenerative brain disease in mammals, including humans.

Viruses are quintessential parasites; they depend on the host cell for almost all of their life-sustaining functions. Unlike true organisms, viruses cannot synthesize proteins, because they lack ribosomes (cell organelles) for the translation of viral messenger RNA (mRNA; a complementary copy of the nucleic acid of the nucleus that associates with ribosomes and directs protein synthesis) into proteins. Viruses must use the ribosomes of their host cells to translate viral mRNA into viral proteins.

Viruses are also energy parasites; unlike cells, they cannot generate or store energy in the form of adenosine triphosphate (ATP). The virus derives energy, as well as all other metabolic functions, from the host cell. The invading virus uses the nucleotides and amino acids of the host cell to synthesize its nucleic acids and proteins, respectively. Some viruses use the lipids and sugar chains of the host cell to form their membranes and glycoproteins (proteins linked to short polymers consisting of several sugars).

The true infectious part of any virus is its nucleic acid, either DNA or RNA but never both. In many viruses, but not all, the nucleic acid alone, stripped of its capsid, can infect (transfect) cells, although considerably less efficiently than can the intact virions.

The virion capsid has three functions: (1) to protect the viral nucleic acid from digestion by certain enzymes (nucleases), (2) to furnish sites on its surface that recognize and attach (adsorb) the virion to receptors on the surface of the host cell, and, in some viruses, (3) to provide proteins that form part of a specialized component that enables the virion to penetrate through the cell surface membrane or, in special cases, to inject the infectious nucleic acid into the interior of the host cell.

Host Range and Distribution

Logic originally dictated that viruses be identified on the basis of the host they infect. This is justified in many cases but not in others, and the host range and distribution of viruses are only one criterion for their classification. It is still traditional to divide viruses into three categories: those that infect animals, plants, or bacteria.

Virtually all plant viruses are transmitted by insects or other organisms (vectors) that feed on plants. The hosts of animal viruses vary from protozoans (single-celled animal organisms) to humans. Many viruses infect either invertebrate animals or vertebrates, and some infect both. Certain viruses that cause serious diseases of animals and humans are carried by arthropods. These vector-borne viruses multiply in both the invertebrate vector and the vertebrate host.

Certain viruses are limited in their host range to the various orders of vertebrates. Some viruses appear to be adapted for growth only in ectothermic vertebrates (animals commonly referred to as cold-blooded, such as fishes and reptiles), possibly because they can reproduce only at low temperatures. Other viruses are limited in their host range to endothermic vertebrates (animals commonly referred to as warm-blooded, such as mammals).

Size and Shape

The amount and arrangement of the proteins and nucleic acid of viruses determine their size and shape. The nucleic acid and proteins of each class of viruses assemble themselves into a structure called a nucleoprotein, or nucleocapsid. Some viruses have more than one layer of protein surrounding the nucleic acid; still others have a lipoprotein membrane (called an envelope), derived from the membrane of the host cell, that surrounds the nucleocapsid core. Penetrating the membrane are additional proteins that determine the specificity of the virus to host cells. The protein and nucleic acid constituents have properties unique for each class of virus; when assembled, they determine the size and shape of the virus for that specific class. The genomes of Mimiviruses and Pandoraviruses, which are some of the largest known viruses, range from 1 to 2.5 Mb (1 Mb = 1,000,000 base pairs of DNA).

Most viruses vary in diameter from 20 nanometres (nm; 0.0000008 inch) to 250–400 nm; the largest, however, measure about 500 nm in diameter and are about 700–1,000 nm in length. Only the largest and most complex viruses can be seen under the light microscope at the highest resolution. Any determination of the size of a virus also must take into account its shape, since different classes of viruses have distinctive shapes.

Shapes of viruses are predominantly of two kinds: rods, or filaments, so called because of the linear array of the nucleic acid and the protein subunits; and spheres, which are actually 20-sided (icosahedral) polygons. Most plant viruses are small and are either filaments or polygons, as are many bacterial viruses. The larger and more-complex bacteriophages, however, contain as their genetic information double-stranded DNA and combine both filamentous and polygonal shapes. The classic T4 bacteriophage is composed of a polygonal head, which contains the DNA genome and a special-function rod-shaped tail of long fibres. Structures such as these are unique to the bacteriophages.

Animal viruses exhibit extreme variation in size and shape. The smallest animal viruses belong to the families Parvoviridae and Picornaviridae and measure about 20 nm and about 30 nm in diameter, respectively. Viruses of these two families are icosahedrons and contain nucleic acids with limited genetic information. Viruses of the family Poxviridae are about 250 to 400 nm in their longest dimension, and they are neither polygons nor filaments. Poxviruses are structurally more complex than simple bacteria, despite their close resemblance. Animal viruses that have rod-shaped (helical) nucleocapsids are those enclosed in an envelope; these viruses are found in the families Paramyxoviridae, Orthomyxoviridae, Coronaviridae, and Rhabdoviridae. Not all enveloped viruses contain helical nucleocapsids, however; those of the families Herpesviridae, Retroviridae, and Togaviridae have polygonal nucleocapsids. Most enveloped viruses appear to be spherical, although the rhabdoviruses are elongated cylinders.

The criteria used for classifying viruses into families and genera are primarily based on three structural considerations: (1) the type and size of their nucleic acid, (2) the shape and size of the capsids, and (3) the presence of a lipid envelope, derived from the host cell, surrounding the viral nucleocapsid.

The Nucleic Acid

As is true in all forms of life, the nucleic acid of each virus encodes the genetic information for the synthesis of all proteins. In almost all free-living organisms, the genetic information is in the form of double-stranded DNA arranged as a spiral lattice joined at the bases along the length of the molecule (a double helix). In viruses, however, genetic information can come in a variety of forms, including single-stranded or double-stranded DNA or RNA.

The nucleic acids of virions are arranged into genomes. All double-stranded DNA viruses consist of a single large molecule, whereas most double-stranded RNA viruses have segmented genomes, with each segment usually representing a single gene that encodes the information for synthesizing a single protein. Viruses with single-stranded genomic DNA are usually small, with limited genetic information. Some single-stranded DNA viruses are composed of two populations of virions, each consisting of complementary single-stranded DNA of polarity opposite to that of the other.

The virions of most plant viruses and many animal and bacterial viruses are composed of single-stranded RNA. In most of these viruses, the genomic RNA is termed a positive strand because the genomic RNA acts as mRNA for direct synthesis (translation) of viral protein. Several large families of animal viruses, and one that includes both plant and animal viruses (the Rhabdoviridae), however, contain genomic single-stranded RNA, termed a negative strand, which is complementary to mRNA. All of these negative-strand RNA viruses have an enzyme, called an RNA-dependent RNA polymerase (transcriptase), which must first catalyze the synthesis of complementary mRNA from the virion genomic RNA before viral protein synthesis can occur. These variations in the nucleic acids of viruses form one central criterion for classification of all viruses.

A distinctive large family of single-stranded RNA viruses is called Retroviridae; the RNA of these viruses is positive, but the viruses are equipped with an enzyme, called a reverse transcriptase, that copies the single-stranded RNA to form double-stranded DNA.

The Protein Capsid

The protein capsid provides the second major criterion for the classification of viruses. The capsid surrounds the virus and is composed of a finite number of protein subunits known as capsomeres, which usually associate with, or are found close to, the virion nucleic acid.

There are two major classes of viruses based on the protein capsid: (1) those in which a single (or segmented) linear nucleic acid molecule with two free ends is essentially completely extended or somewhat coiled (a helix) and (2) those in which the nucleic acid, which may or may not be a covalently closed circle, is wound tightly into a condensed configuration, like a ball of yarn. These two classes of virus assume in the first case a long, extended rodlike structure and in the second case a symmetrical polygon.

By far the best-studied example of a helical rod-shaped virus is the tobacco mosaic virus, which was crystallized by Wendell Stanley in 1935. The tobacco mosaic virus contains a genome of single-stranded RNA encased by 2,130 molecules of a single protein; there are $16^1/_3$ protein molecules for each turn of the RNA helix in the ratio of three nucleotides for each protein molecule.

Under the right environmental conditions, viral RNA and protein molecules in liquid suspension will assemble themselves into a perfectly formed and fully infectious virus. The length of the helical virus capsid is determined by the length of the nucleic acid molecule, which is the framework for the assembly of the capsid protein. The various helical viruses have different lengths and widths, depending on the size of the nucleic acid as well as on the mass and shape of the protein molecule. Some of these helical viruses form rigid rods, while others form flexible rods, depending on the properties of the assembled proteins.

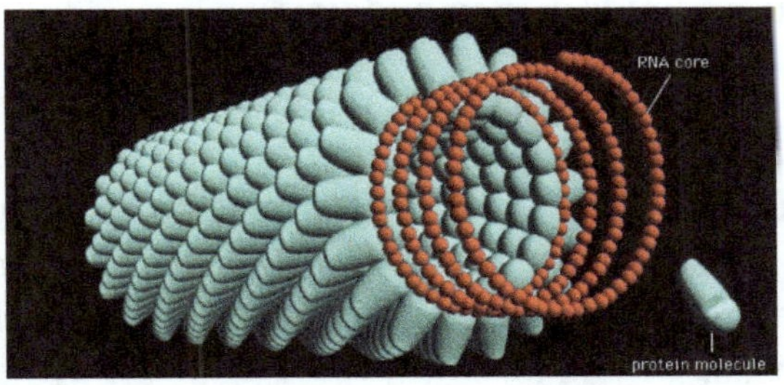

Schematic structure of the tobacco mosaic virus: The cutaway section shows the helical ribonucleic acid associated with protein molecules in a ratio of three nucleotides per protein molecule.

Polygonal viruses vary greatly in size, from 20 to 150 nm in diameter, essentially proportional to the size of the nucleic acid molecule coiled up inside the virion. Most, if not all, of the polygonal viruses are icosahedral; like a geodesic dome, they are formed by equilateral triangles, in this case 20. Each triangle is composed of protein subunits (capsomeres), often in the form of hexons (multiples of six) that are the building blocks of the capsid. There are 12 vertices (the apical junctions of these 20 triangles), each comprising a penton (five subunits). These icosahedral virions have three axes of fivefold, threefold, and twofold rotational symmetry. The number of capsomeres is a basis for taxonomic classification of these virus families. Certain icosahedral viruses, usually those that are more complex, contain internal proteins adhering to the nucleic acid that are not accessible at the surface of the virions.

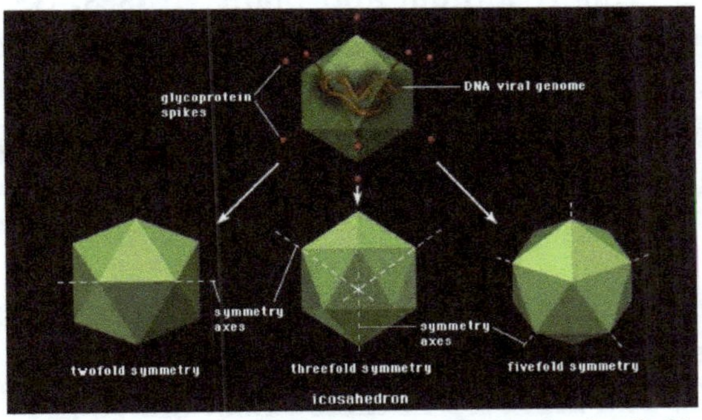

A virus icosahedron (20-sided structure) shown in the (left) twofold, (centre) threefold, and (right) fivefold axes of symmetry. Edges of the upper and lower surfaces are drawn in solid and broken lines, respectively.

Lipoprotein Envelope

Surrounding viruses of either helical or icosahedral symmetry are lipoprotein envelopes, unit membranes of two lipid layers interspersed with protein molecules (lipoprotein bilayer). These viral membranes are composed of phospholipids and neutral lipids (largely cholesterol) derived from cell membranes during the process known as budding. Virtually all proteins of the cell membrane, however, are replaced by proteins of viral origin during budding. Although all the viral envelope lipids originate from the cell, their relative proportions vary from those in the cell membrane because the viral proteins select only certain lipids during budding.

Associated with the virion membrane are "integral" glycoproteins, which completely traverse the lipid bilayer, and "peripheral" matrix proteins, which line the inner surface. The glycoproteins contain regions of amino acids that, in the first step of viral infection, recognize host-cell receptors. Matrix proteins appear to function in the selection of regions of the cell membrane to be used for the viral membrane, as well as to aid the virion in entering cells.

Cycle of Infection

Viruses can reproduce only within a host cell. The parental virus (virion) gives rise to numerous progeny, usually genetically and structurally identical to the parent virus. The actions of the virus depend both on its destructive tendencies toward a specific host cell and on environmental conditions. In the vegetative cycle of viral infection, multiplication of progeny viruses can be rapid. This cycle of infection often results in the death of the cell and the release of many virus progeny. Certain viruses, particularly bacteriophages, are called temperate (or latent) because the infection does not immediately result in cell death. The viral genetic material remains dormant or is actually integrated into the genome of the host cell. Cells infected with temperate viruses are called lysogenic because the cells tend to be broken down when they encounter some chemical or physical factor, such as ultraviolet light. In addition, many animal and plant viruses, the genetic information of which is not integrated into the host DNA, may lie dormant in tissues for long periods of time without causing much, if any, tissue damage. Viral infection does not always result in cell death or tissue injury; in fact, most viruses lie dormant in tissue without ever causing pathological effects, or they do so only under other, often environmental, provocations.

Although the reproductive pathways of different viruses vary considerably, there are certain basic principles and a particular series of events in the cycle of infection for most, if not all, viruses. The first step in the cycle of infection is that the invading parental virus (virion) must attach to the surface of the host cell (adsorption). In the second step, the intact virion either penetrates the outer membrane and enters the cell's interior (cytoplasm) or injects the genetic material of the virus into the interior of the cell while the protein capsid (and envelope, if present) remains at the cell surface. In the case of whole-virion penetration, a subsequent process (uncoating) liberates the genetic material from the capsid and envelope, if present. In either case, the viral genetic material cannot begin to synthesize protein until it has emerged from the capsid or envelope.

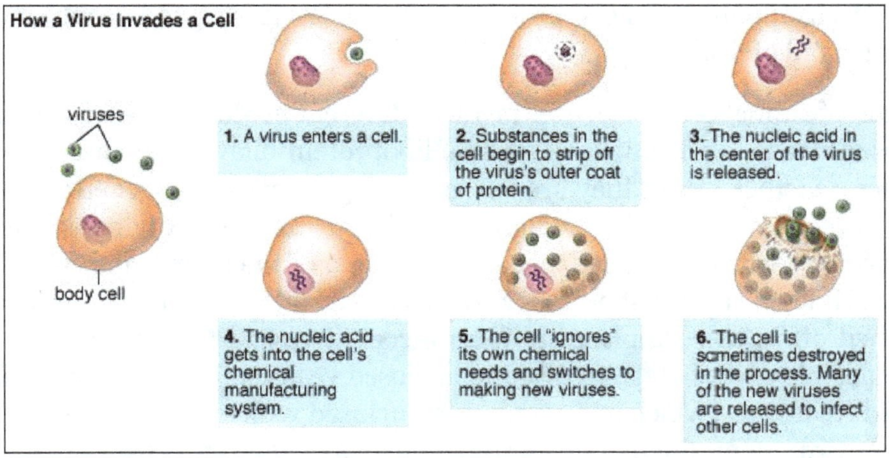

Virus: Invasion of a cell.

Certain bacterial viruses, such as the T4 bacteriophage, have evolved an elaborate process of infection: following adsorption and firm attachment of the virus's tail to the bacterium surface by means of proteinaceous "pins," the musclelike tail contracts, and the tail plug penetrates the cell wall and underlying membrane and injects virus (phage) DNA into the cell. Other bacteriophages penetrate the cell membrane by different means, such as injecting the nucleic acid through the male (sex) pili of the bacterium. In all bacterial viruses, penetration transmits the viral nucleic acid through a rigid bacterial cell wall.

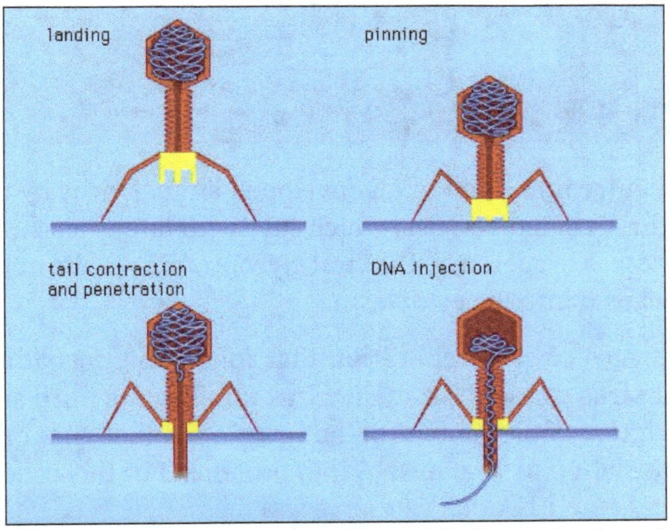

General structure of T4 bacteriophage and a model of its mode of attachment to, and injection of its DNA into, a bacterial cell.

Plant cells also have rigid cell walls, which plant viruses cannot ordinarily penetrate. Plant viruses, however, have not evolved their own systems for injecting nucleic acids into host cells, and so they are transmitted by the proboscis of insects that feed on plants. In the laboratory, plant viruses penetrate plant cells if the cell walls have been abraded with sandpaper or if cell protoplasts (plasma membrane, cytoplasm, and nucleus) are devoid of walls.

Penetration of animal cells by viruses involves different processes, because animal cells are enclosed not by walls but by a flexible lipoprotein bilayer membrane. Most animal viruses, whether or not they are encased in lipid envelopes, penetrate cells in an intact form by a process called endocytosis. The membrane invaginates and engulfs a virus particle adsorbed to a cell, usually in an area of the membrane called a coated pit, which is lined by a special protein known as clathrin. As the coated pit invaginates, it is pinched off in the cytoplasm to form a coated vesicle. The coated vesicle fuses with cytoplasmic endosomes (membrane-enclosed vesicles) and then with cell organelles called lysosomes, which are membrane-enclosed vesicles containing enzymes. In an acidic environment, the membrane of an enveloped virus fuses with the endosome membrane, and the viral nucleocapsid is released into the cytoplasm. Nonenveloped viruses presumably undergo a similar process, by which the protein capsid is degraded, releasing the naked viral nucleic acid into the cytoplasm.

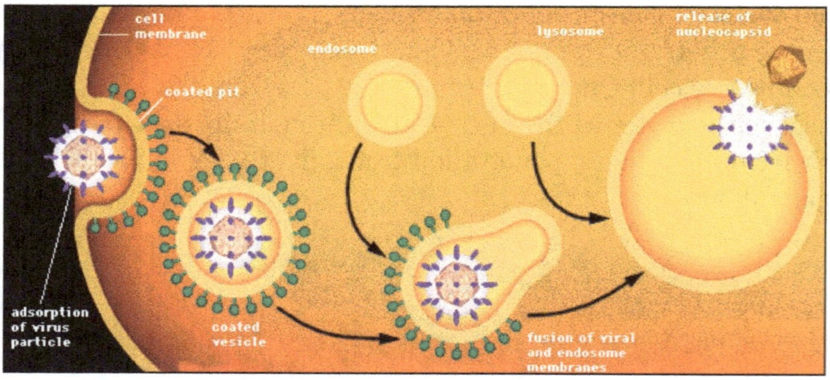

Adsorption to and entry into a cell of an enveloped animal virus by the process of endocytosis into clathrin-coated vesicles, which fuse with large vesicles (endosomes and lysosomes). The process triggered by the viral glycoprotein results in fusion and release of the viral nucleocapsid into the cytoplasm.

The order of the stages of viral replication that follow the uncoating of the genome varies for different virus classes. For many virus families the third step in the cycle of infection is transcription of the genome of the virus to produce viral mRNA, followed by the fourth step, translation of viral mRNA into proteins. For those viruses in which the genomic nucleic acid is an RNA that can serve as a messenger (i.e., positive-strand RNA viruses), the third step is the translation of the RNA to form viral proteins; some of these newly synthesized viral proteins are enzymes that synthesize nucleic acids (polymerases), which carry out a fourth step, the transcription of more mRNA from the viral genome. For the more complicated DNA viruses, such as adenoviruses and herpesviruses, some regions of the genome synthesize "early" mRNAs, which are translated into polymerases that initiate the transcription of "late" regions of the DNA into mRNAs, which are then translated into structural proteins.

Regardless of how the third and fourth steps proceed, the fifth step in the cycle of infection is replication (reproduction of the parental genome to make progeny genomes). The sixth step is the assembly of the newly replicated progeny genomes with structural proteins to make fully formed progeny virions. The seventh and last step is the release of progeny virions by lysis of the host cell through the process of either extrusion or budding, depending on the nature of the virus. In a host animal or cell culture, this seven-step process may be repeated many times; the progeny virions released from the original site of infection are then transmitted to other sites or to other individuals.

For most animal and plant RNA viruses, all replicative events take place in the cytoplasm; in fact, many of these RNA viruses can grow in host cells in which the nucleus has been removed. Replication of most animal and plant DNA viruses, as well as the RNA influenza virus, takes place in the nucleus. In these viruses, transcription takes place in the nucleus, the mRNA migrates to the cytoplasm, where it is translated, and these viral proteins migrate back to the nucleus, where they assemble with newly replicated progeny genomes. Migration of newly translated viral proteins from the cytoplasm to the nucleus is generally a function of specific amino acid sequences called "signals," which translocate the protein through pores in the nucleus membrane.

Viral DNA Integration

Lysogeny

Many bacterial and animal viruses lie dormant in the infected cell, and their DNA may be integrated into the DNA of the host cell chromosome. The integrated viral DNA replicates as the cell genome replicates; after cell division, the integrated viral DNA is duplicated and usually distributed equally to the two cells that result. The bacteria that carry the noninfective precursor phage, called the prophage, remain healthy and continue to grow until they are stimulated by some perturbing factor, such as ultraviolet light. The prophage DNA is then excised from the bacterial chromosome, and the phage replicates, producing many progeny phages and lysing the host bacterial cell. This process, originally discovered in temperate bacteriophages in 1950 by the French microbiologist André Lwoff, is called lysogeny.

The classic example of a temperate bacteriophage is called lambda (λ) virus, which readily causes lysogeny in certain species of the bacterium Escherichia coli. The DNA of the λ bacteriophage is integrated into the DNA of the E. coli host chromosome at specific regions called attachment sites. The integrated prophage is the inherited, noninfectious form of the virus; it contains a gene that represses the lytic functions of the phage and thus ensures that the host cell will continue to replicate the phage DNA along with its own and that it will not be destroyed by the virus. Ultraviolet light, or other factors that stimulate the replication of DNA in the host cell, causes the formation of a recA protease, an enzyme that breaks apart the λ phage repressor and induces λ phage replication and, eventually, destruction of the host cell.

Excision of the prophage DNA from the host chromosomal DNA (as an initial step in the synthesis of an infective, lytic virus) sometimes results in the removal of some of the host cell DNA, which is packaged into defective bacteriophages; part of the bacteriophage DNA is removed and replaced at the other end by a gene of the host bacterium. Such a virus particle is called a transducing phage because, when it infects a bacterial cell, it can transmit the gene captured by λ phage DNA into the next bacterial cell it infects. Transduction by bacteriophages is an efficient means for transferring the genetic information of one bacterial cell to another.

This means of transferring genetic information, called lysogenic conversion, imparts genes with special functions to bacterial cells without such functions. It is common in bacteria and is an important aspect of the epidemiology (incidence, distribution, and control) of infectious diseases. For example, the bacterium Coryne bacterium diphtheriae is the causative agent of diphtheria, but only when it contains the prophage of bacteriophage β, which codes for the toxin that is responsible for the disease.

Malignant Transformation

A phenomenon analogous to bacterial cell lysogeny occurs in animal cells infected with certain viruses. These animal viruses do not generally cause disease immediately for certain animal cells. Instead, animal cells are persistently infected with such viruses, the DNA of which (provirus) is integrated into the chromosomal DNA of the host cell. In general, cells with integrated proviral DNA are converted into cancer cells, a phenomenon known as malignant transformation. As is the case with bacterial prophages, the transformed animal cell contains no infectious virus but only the integrated provirus DNA, which replicates along with the dividing cell's chromosomes. Therefore, following mitosis of the transformed cell, each new cell receives a copy of the proviral DNA. The hallmark of these transformed animal cells is that their growth is uncontrollable; unlike normal cells, their growth is not inhibited by contact with other cells, and they lose their capacity to adhere (anchor) to certain surfaces. Growth of normal tissues and organs is also controlled by a genetic phenomenon called programmed cell death, or apoptosis, in which a certain number of cells will die and be eliminated after a finite number of divisions. Malignant transformation can impede programmed cell death, thus allowing the cells to grow uncontrolled and resulting in cancer.

Among the animal viruses that cause malignant transformation by integration of proviral DNA are several families of DNA viruses and one large family of RNA viruses, the Retroviridae. Viruses of the family Polyomaviridae, a group of papovaviruses, were perhaps the first to be associated with malignancy (causing death or illness) in animals. Polyomaviruses are widespread in mice; they can infect other rodents, and they can cause tumours in infected animals. Another virus of the family Polyomaviridae is simian virus 40 (SV40), originally isolated from cells of the African green monkey (Cercopithecus sabaeus), where it grows rapidly and kills the cells. Infection of rodent or human cells, however, results in an abortive infection (an incompatibility between the virus and the host cell) but sometimes induces malignancy (sarcomas or lymphomas) in the occasional cell that is transformed. Viruses related to polyomavirus and SV40 have been isolated from humans, one of which, the JC virus, appears to be the causative agent of a fatal neurological disease called progressive multifocal leukoencephalopathy. In general, however, the human papovaviruses are not clearly associated with disease.

Other papovaviruses include the papillomaviruses (family Papillomaviridae), which are also small polygonal viruses containing circular double-stranded DNA. The papillomaviruses are associated with usually benign (nonthreatening) but widespread tumours, called papillomas or polyps, occurring in human skin and the genital tract. Specific papillomaviruses have been identified in humans in common warts and in genital warts (condylomata acuminata). Cancers of the human genital tract, particularly uterine cancer of the cervix, are frequently found in association with human papillomavirus type 16 (HPV 16); the virus undoubtedly is transmitted as a venereal disease.

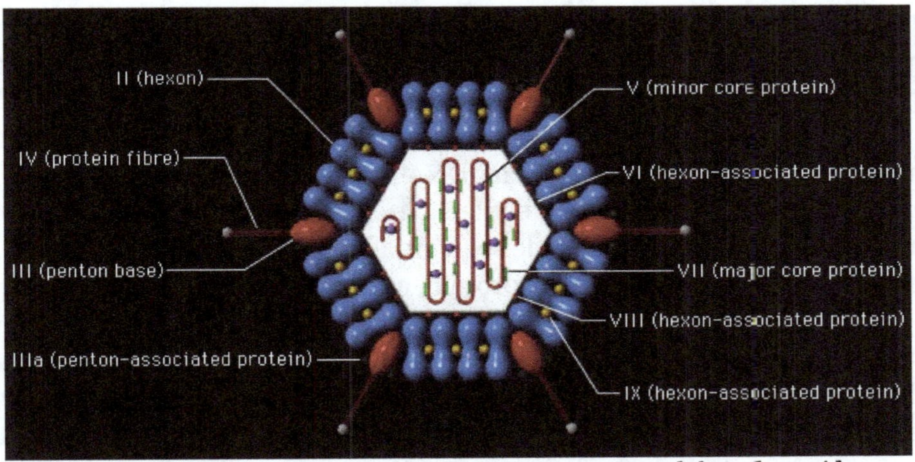

The structure of an adenovirus showing its components and the polypeptides (proteins) exhibited by staining after electrophoresis.

Certain viruses of the family Adenoviridae, originally found in the tonsils and adenoids of humans, cause malignant transformation in certain cells. This phenomenon of cancer induction under laboratory conditions has been studied widely, but there is no evidence that the common adenoviruses cause cancers in humans. The common viruses of the family Herpesviridae, however, including the common herpes simplex viruses that cause cold sores and the venereal disease genital herpes, are suspected of being causative agents of cancer. Like the adenoviruses, the herpesviruses can cause malignant transformations, and their DNA is integrated into the host cell chromosome. A herpesvirus known as the Epstein-Barr virus causes a frequently fatal childhood cancer called Burkitt lymphoma as well as the nonmalignant disease infectious mononucleosis. The herpesvirus cytomegalovirus lies dormant in the tissues of most humans and can be induced to cause fatal diseases in infants and immunocompromised adults. A different herpesvirus causes chickenpox (varicella); the same virus lies latent in the tissues for long periods of time (perhaps years or decades) and later undergoes recrudescence (the recurrence of symptoms after they have abated) to cause the painful skin and neurological disease called herpes zoster, or shingles. In addition, there are herpesviruses that cause disease in animals—for example, the widespread and usually fatal disease in chickens called Marek's disease. The widespread distribution of viruses of the family Herpesviridae is evident from other diseases in monkeys and frogs.

The viruses of the family Retroviridae are perhaps the most widely distributed of the transforming viruses that infect eukaryotic cells ranging from yeast to humans. It was suggested early in the 20th century that viruses cause leukemias and lymphomas in birds. In 1911 the American pathologist Peyton Rous first described a virus that causes sarcomas in chickens.

The virions of retroviruses are spherical (or polygonal) and are surrounded by a lipid membrane containing a glycoprotein that recognizes and binds to cell receptors of a particular species (type-specific glycoproteins). Retrovirus genomes consist of two identical RNA molecules, each with 7,000 to 10,000 nucleotides. Associated with the virion RNA is an enzyme, an RNA-dependent DNA polymerase, also called a reverse transcriptase. Using the virion RNA as a template, the reverse transcriptase catalyzes the synthesis of a linear DNA molecule complementary to the virion RNA. The new complementary strand of DNA also serves as a template for the reverse transcriptase, which makes a second anticomplementary DNA molecule, thus forming double-stranded DNA. The genomic RNA of fully infectious bird retroviruses, those that can replicate autonomously, has four genes that code sequentially for group-specific antigens, the reverse transcriptase, the envelope glycoprotein, and the sarcoma-transforming protein. At each end of the genome are homologous flanking nucleotide sequences, known as long terminal repeats (LTR), which code for double-stranded DNA that can recognize host cell DNA sequences for integration of the proviral DNA into the host cell chromosome. Many retroviruses are defective and cannot replicate in cells without helper (nondefective) retroviruses. The helper retroviruses generally transform fibroblastic cells, resulting in malignant sarcomas, whereas the defective retroviruses transform blood-cell precursors, resulting in leukemias.

Many different retroviruses have been identified as causative agents of cancers in birds, rodents (particularly mice), domestic cats, monkeys, and humans. Certain lymphatic leukemias in humans are caused by human T-cell leukemia virus (HTLV); acquired immune deficiency syndrome (AIDS) is caused by a retrovirus called human immunodeficiency virus (HIV).

Retroviruses originated from genes in many different species of animals and even lower forms of life. Individual retroviruses are limited in their host range and do not readily cross species barriers. Virtually every retrovirus studied to date is analogous to the genes normally found in animals (including humans), known as proto-oncogenes, genes that are involved with regulating normal cell growth and development and that also have the potential to change into cancer-causing genes. These proto-oncogenes have deoxynucleotide sequences closely, but not entirely, homologous (i.e., of the same type and order) to the nucleotide sequences of a corresponding viral cancer-causing gene, called an oncogene. Integration of retrovirus DNA into cell chromosomes results in cancer, but the proto-oncogenes do not become cancer-causing genes unless triggered by another event. Cancers caused by chemical or physical carcinogens in the environment probably often, if not invariably, are due to alterations in the sequences of proto-oncogenes that have converted them to oncogenes. Some of the DNA tumour viruses, such as SV40 or adenoviruses, may induce malignant transformation when their DNA is integrated in proximity to the site of a proto-oncogene. All cancers studied to date appear to be due to either mutations in proto-oncogenes or the inheritance of mutated tumour suppressor genes, which normally regulate the function of proto-oncogenes.

Virology

Virology is the study of viruses and virus-like agents, including (but not limited to) their taxonomy, disease-producing properties, cultivation and genetics. It is often considered a part of microbiology or pathology. In the early years this discipline was dependent upon advances in the chemical and physical sciences, but viruses soon became tools for probing basic biochemical processes of cells.

Branches of Virology

2

- Clinical Virology
- Synthetic Virology
- Neurovirology
- Molecular Virology
- Veterinary Virology
- Paleovirology

There are different branches of virology which can be classified as clinical virology, synthetic virology, neurovirology, veterinary virology, paleovirology and molecular virology. This chapter has been carefully written to provide an easy understanding of the various branches of virology.

Clinical Virology

Clinical Virology is a field of medicine and virology that is concerned with the study of viruses that cause human pathologies. The clinical virology laboratory stands to benefit greatly from LCR because of the difficulty and expense associated with viral culture. Confirmation of viral infection and resolution of indeterminate viral serology are important applications of LCR.

Human Immunodeficiency Virus

All blood donations are currently screened for human immunodeficiency viruses (HIV), the human T-lymphotropic virus type I (HTLV-1), and the hepatitis C virus (HCV) using enzyme immunoassay (EIA) methods. Seropositive results are confirmed by Western blots or radioimmunoprecipitation assays (RIPA). Depending on the viral marker, indeterminate confirmatory results can occur in 20–35% of the EIA positive specimens. Ideally, a confirmatory assay should provide direct rather than indirect evidence of viral infection. Because of its high sensitivity and specificity, LCR can be a valuable tool for the direct detection of viral infections. In addition, LCR can be used to clarify the serological status of individuals with indeterminate confirmatory test results.

Quantitative LCR for HIV, hepatitis B virus (HBV), or HCV may also be useful in monitoring therapeutic efficacy or disease progression. The quantitative aspect of LCR amplification is illustrated in figure below. Dilutions of a plasmid containing HIV-1 DNA were detected using LCR oligonucleotides specific for a portion of the pol gene. LCR probes can be designed for other proviral genes, for genomic RNA, or for both to help establish the clinical significance of changes in copy numbers during therapy.

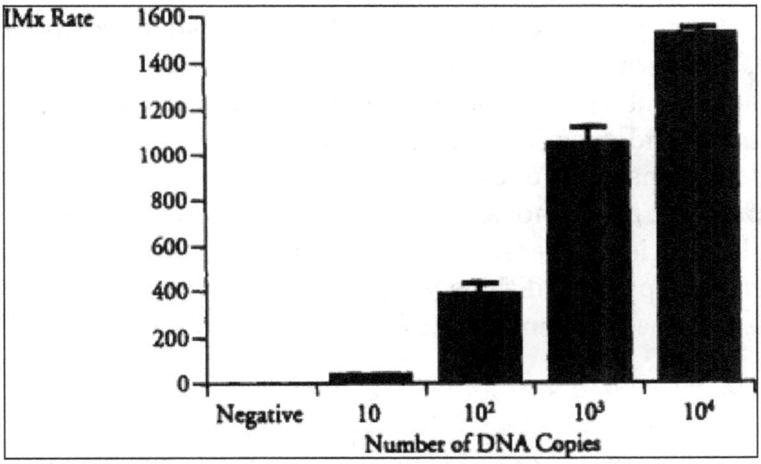

Histogram showing LCR results for the detection of different copy numbers of target HIV proviral DNA.

Human Papillomavirus

Because most microorganisms contain both highly conserved, genus-specific DNA sequences and strain- or species-specific sequences, the appropriate target sequence can allow the user to detect a group of organisms or to differentiate species or subtypes. The importance of viral typing is illustrated by the fact that more than 60 human papillomavirus (HPV) types have been detected and at least 20 HPV types have been detected in the human genital tract. In addition, several HPV types are associated with the development of cervical cancer. HPV typing of women with cytologically abnormal Pap smears can contribute to the clinical management and follow-up of these individuals. Due to the specificity of LCR, it is possible to develop assays that can distinguish between different HPV types. In one study, an LCR assay specific for HPV type 16 (HPV-16) was used to correctly identify 16 of 16 tissue samples determined to be positive for HPV-16 by Southern blot. Further, a second LCR assay specific for HPV type 18 (HPV-18) was used to correctly identify 5 of 5 tissue samples determined to be positive for HPV-18 by Southern blot.

Human T-Lymphotrophic Virus Type I and Type II

Another application of LCR for viral typing is the differentiation of HTLV-I from HTLV-II, the former being the cause of adult T-cell leukemia/lymphoma and a progressive neurological disorder similar to multiple sclerosis, whereas the latter virus is not yet associated with any disease. Because of a high degree of amino acid homology, differentiation between the two viruses is difficult by serology. Identification and follow-up of individuals with HTLV-II infection may help define its role in pathogenesis.

Chlamydia Trachomatis and Neisseria Gonorrhoeae

In addition to HPV typing, diagnostic applications for LCR in the area of sexually transmitted disease include the detection of Chlamydia trachomatis, Neisseria gonorrhoaeae, and herpes simplex virus (HSV). The isolation of C. trachomatis and N. gonorrhoeae by culture remains the "gold standard" since culture provides the highest sensitivity and specificity compared with antigen or antibody testing by EIA. However, the sensitivity of culture is estimated to be 75–85% for C. trachomatis and 85–95% for N. gonorrhoaeae. Furthermore, culture sensitivities vary greatly in different laboratories. Several factors such as the use of antibiotics by the patient, specimen transport conditions, and transport time affect culture efficiencies. These obstacles can be overcome by LCR because the organisms need not be viable for detection.

The detection of C. trachomatis by LCR requires that the selected target sequences be present in all variants but absent from other microorganisms, particularly those commonly found in genital specimens. LCR detection of all 15 recognized C. trachomatis serological variants (serovars) has been demonstrated but the C. trachomatis target sequence was not detected in 88 other potentially cross-reacting microorganisms. The cutoff values are validated by rigorous analysis of clinical specimens, concentrating on those giving discordant results between culture and LCR, so that optimal sensitivity and specificity are achieved. Chlamydia trachomatis studies by LCR on 2003 female endocervical specimens demonstrated 94.7% sensitivity and 99.8% specificity compared with 70.8% sensitivity and 100% specificity for culture. LCR detection of N. gonorrhoeae in 2235 endocervical specimens resulted in 98.9% sensitivity and 99.6% specificity compared with 88.9% sensitivity and 100% specificity (by definition) for culture.

Asymptomatic male carriers represent a reservoir population largely untested for C. trachomatis infection because of the difficulty of obtaining urethral swab specimens in males. Culture of urine specimens for C. trachomatis is usually not practical because of low sensitivity and the variable results caused by the cytopathic effects of urine. In addition, the number of organisms is often lower in urine than in urogenital swab specimens. As a result, an easy and sensitive LCR procedure for the detection of C. trachomatis in urine was developed. The performance of C. trachomatis detection by LCR on male urine specimens compared with culture on urethral swab specimens from the same individuals. The results of 496 paired specimens demonstrated 95.7% (45/47) sensitivity for LCR compared with 44.7% (21/47) for culture on urethral swabs, with LCR detecting 51.0% more positives than culture. Of the 26 discordant specimen pairs that were positive by LCR but negative by culture, 24 (79.2%) were confirmed as true positives by direct fluorescent antibody (DFA) analysis of a cytospin sediment from the culture specimen, and the remaining 2 (20.8%) discordant specimen pairs were confirmed by LCR using probes directed against the gene encoding the major outer membrane protein (MOMP) of C. trachomatis. The screening of urine by LCR should have a positive impact on the control of C. trachomatis infections, particularly with respect to the identification of asymptomatic male carriers.

Table: Comparison of chlamydia trachomatis detection by lcr on male urine specimens with culture on matched urethral specimens.

	Urethral culture	
Urine LCR	Positive	Negative
Positive	19	26
Negative	2	449
a: Chlamydia trachomatis LCR sensitivity: 95.7% (45/47); Chlamydia trachomatis LCR specificity: 100% (449/449); Chlamydia trachomatis culture sensitivity: 44.7% (21/47). b: All 26 confirmed as positive.		

Application of molecular biology techniques to clinical virology has introduced significant changes in enteroviral diagnosis. Sensitivity and specificity has increased significantly. In addition, molecular techniques provide short detection times, detection of isolates that do not readily grow in cell culture, and allow typing of enteroviruses that are difficult to identify by traditional techniques.

Nucleic acid probe hybridization tests were used to detect enteroviruses, but have been replaced by RT-PCR assays. In general, diagnostic assays for enterovirus are targeted to amplify regions of the 5' NTR.

The availability of sequence data for all members of the genus has enabled differentiation of viruses based on the nucleic acid sequence encoding VP1. Molecular identification can be performed using sequences of RT-PCR products obtained directly from clinical specimens or virus isolates. Although other regions have been used for molecular typing they fail to provide reliable identification. Only VP1 capsid sequence correlates with serotype, due to the high frequency of interspecies recombination among co-circulating enteroviruses (e.g. within HEV-B). Multiple new enteroviruses have been detected and characterized using molecular typing and sequencing. Phylogenetic studies of clinical isolates are useful in examining enteroviral evolution and to characterize enterovirus molecular epidemiology.

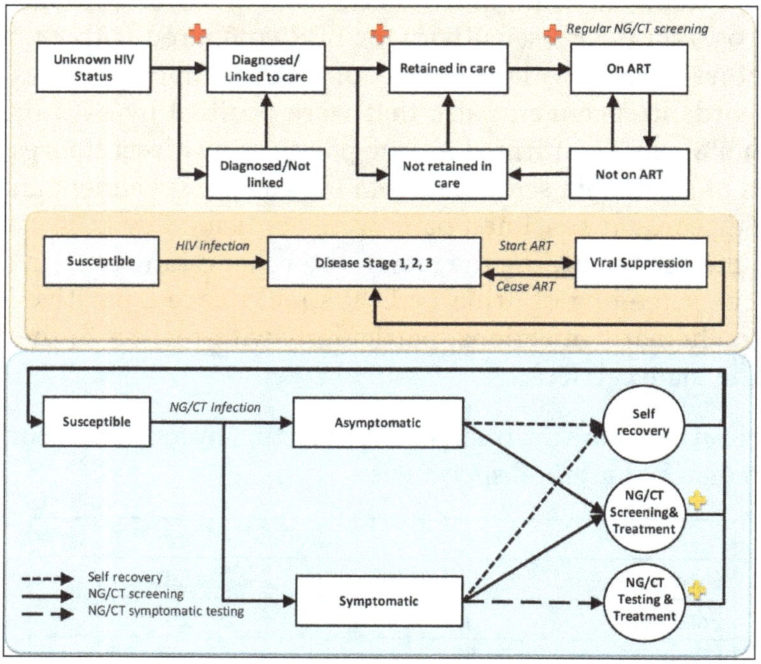

Human Enteroviruses

Detection and Typing

Application of molecular biology techniques to clinical virology has introduced significant changes in enteroviral diagnosis. Reliability has increased significantly. In addition, molecular techniques provide short detection times and diagnosis of isolates that do not readily grow in cell culture, and allow typing of enteroviruses that are difficult or impossible to identify by traditional techniques.

Nucleic acid probe hybridization tests were the first application of molecular techniques to detect enteroviruses, but have been replaced by the polymerase chain reaction (PCR). In general, diagnostic assays for enterovirus are targeted to amplify conserved fragments of the 5′ NTR of the virus genome.

The availability of sequence data for all members of the Enterovirus genus has enabled differentiation of viruses based on the nucleic acid sequence encoding the VP1 capsid protein. Molecular identification can be performed using sequences of the reverse transcriptase PCR (RT-PCR) amplification products obtained directly from clinical specimens or virus isolates. Sequences surrounding the VP4–VP2 junction have also been used for molecular typing, but this region does not always provide a reliable identification. Only VP1 capsid sequence correlates with serotype, due to the high frequency of interspecies recombination among co-circulating enteroviruses (e.g. within HEV-B). Several new enteroviruses, including EV73, EV74, EV75, EV77 and EV78, EV79–88, EV97, EV100 and EV101, have been detected and characterized using molecular typing systems and sequencing.

Synthetic Virology

Synthetic virology is a branch of virology engaged specifically in the study and engineering of synthetic, man-made viruses. It is a multidisciplinary research field at the intersection of virology, synthetic biology, computational biology, and DNA nanotechnology, from which it borrows and integrates its concepts and methodologies. There is a wide range of applications for synthetic viral technology such as medical treatments, investigative tools, and reviving organisms.

Constructing de Novo Synthetic Viruses

Advances in genome sequencing technology and oligonucleotide synthesis paved the way for construction of synthetic genomes based on previously sequenced genomes. Both RNA and DNA viruses can be made using existing methods. RNA viruses have historically been utilized due to the typically small genome size and existing reverse transcription machinery present. The first man-made infectious viruses generated without any natural template were of the polio virus and the φX174 bacteriophage. With synthetic live viruses, it is not whole viruses that are synthesized but rather their genome at first, both in the case of DNA and RNA viruses. For many viruses, viral RNA is infectious when introduced into a cell (during infection or after reverse transcription). These organisms are able to sustain an infectious life cycle upon introduction in vivo.

Applications

This technology is now being used to investigate novel vaccine strategies. The ability to synthesize viruses has far-reaching consequences, since viruses can no longer be regarded as extinct, as long as the information of their genome sequence is known and permissive cells are available. As of March 2014, the full-length genome sequences of 3,843 different viruses, including the smallpox virus, are publicly available in an online database maintained by the National Institutes of Health. Synthetic viruses have also been researched as potential gene therapy tools.

Neurovirology

Neurovirology is an interdisciplinary field which represents a melding of clinical neuroscience, virology, immunology, and molecular biology. The main focus of the field is to study viruses capable of infecting the nervous system. In addition to this, the field studies the use of viruses to trace neuroanatomical pathways, for gene therapy, and to eliminate detrimental populations of neural cells.

The field of neurovirology was formed within the past 30 years. It was founded upon the discovery that a large number of viruses are capable of invading and establishing latent infections in nervous tissue. Such viruses have been shown to produce slow, chronic, or progressive nervous system diseases. Neurovirology incorporates the related fields of virology, neuroscience, neurology, immunology, and molecular biology. The main focus of the field is to study the molecular and biological basis of virus induced diseases of the nervous system. In addition to this, the field studies the use of these viruses as tracers of neuroanatomical pathways and as vectors for gene therapy.

The field relies upon neuroimaging, isolation of the virus from brain tissue or CSF, serological testing of serum and CSF, and microscopic examination of tissue to diagnose nervous system infections.

Neurovirology only became an official field within the past 30 years. However, the true origin of neurovirology can be accredited to the discovery that some viruses may have an affinity for nervous system tissue. This discovery was made in the late 1880s with research involving rabies.

In 1881, while studying rabies, Louis Pasteur demonstrated that the central nervous system played a crucial role in the progression of the disease. Following this discovery, in 1890, Schaffer demonstrated histological evidence that the rabies virus spread via neural networks.

In 1929 Heinrich Pette established the first classification criteria for inflammatory diseases of the nervous system. This classification separated the diseases into two groups: gray matter acute and white matter acute inflammatory diseases. Gray matter acute inflammatory diseases were characterized by damage to neurons with myelin remaining intact. White matter acute inflammatory diseases were characterized by destruction of the myelin, with neurons remaining intact.

In 1938, Sbin and Olitsky discovered that the distribution of the virus within the body depended on its mechanism of entry.

In 1965, ZuRhein and Chou established that destruction of myelin could result from primary virus infection, not only from autoimmune response to the virus.

Most of the research of which the field of neurovirology is based upon occurred in the late 1980s and the 1900s.

Beginning in 1999 the International Society of Neurovirology has recognized and awarded individuals who have contributed significantly to the field with the Pioneer in NeuroVirology Award.

DNA Virus Family

Herpesviruses

- Herpes simplex virus type 1 (HSV-1):
    - Evidence may support link between virus and Alzheimer's disease, intractable focal epilepsy, and acute disseminated encephalomyelitis.
    - Infection can produce myelitis and encephalitis.
- Herpes simplex virus type 2 (HSV-2):
    - Infection can produce myelitis and encephalitis.
- Human cytomegalovirus (CMV):
    - Infection is associated with Guillain–Barré syndrome.
    - Is the most common viral cause of childhood mental retardation.
    - Is linked to neurological abnormalities, such as: microencephaly, seizures, hypotomia, and sensorineural hearing loss.
- Varicella-zoster virus (VZV):
    - Is associated with neurological complications such as: meningoencephalitis, myelitis, cerebellar ataxia, Reye's syndrome, cranial neuropatheis, and postinfectious disseminated encephalomyelitis.
- Epstein-Barr virus (EBV):
    - Is associated with Guillain–Barré syndrome, cranial mononeuropathies, focal encephalitis, and myelitis.
    - Is considered the leading viral candidate for causing multiple sclerosis.
- Human herpesvirus type 6 (HHV-6) and Human herpesvirus type 7 (HHV-7):
    - Have not been linked to human nervous system disease.

Polyomaviruses

- JC virus (JCV):
    - Is associated with progressive multifocal leukoencephalopathy and demyelination.

RNA Virus Family

Rhabdoviruses

- Rabies virus
    - Gives rise to neuronal dysfunction.

Paramyxoviruses

- Measles virus:
    - Is a major cause of neurological deficits.
- Mumps virus:
    - Is the leading cause of virus induced aseptic meningitis and encephalitis

Retroviruses

- Human immunodeficiency virus (HIV):
    - Is associated with cognitive dysfunction.

Viral Entry into the Nervous System

Viruses have evolved mechanisms enabling them to easily infiltrate the nervous system. Two main methods of viral entry have been identified: transneuronal spread and hematogenous spread.

Transneuronal Spread

The mechanism behind transneuronal spread is not entirely known yet, but it involves the virus escaping the immune system by traveling up the axons of the nerves.

Hematogenous Spread

There are two main ways that a virus is thought to enter the brain via hematogenous spread. The first is by infecting an immune cell, which then carries the virus to the nervous tissue. Viral examples of this include the JC virus which infects B cells and HIV which infects CD4 T cells and macrophages to infiltrate the brain. The second is by crossing the blood capillaries as a free virus or in leukocytes.

Advantages of Infecting the Nervous System

Neurons lack molecules necessary to present viral peptides on the surface to killer cells, which means they provide a safe house for viruses to replicate. Once viruses get in neurons they can persist for the hosts lifetime and can influence the factors that disturb the function of neurons and the homeostasis of the nervous system, leading to nervous system diseases.

Tools used for Diagnosing Neuroviral Infections

There are several diagnostic tools which have become invaluable to diagnosing viral infections of the nervous system. In the past, more invasive methods of obtaining samples for diagnosis were needed such as the use of brain biopsy. Now, with the advancement of technology, less invasive means are used more frequently, such as neuroimaging and the analysis of cerebrospinal fluid (CSF).

Neuroimaging

CT scans and MRI scans are useful in visualizing inflammation and lesions caused by viral infection of the CNS. MRI is used to visualize deep white matter and temporal lobe lesions, which are not well defined by a CT scan.

Lumbar Puncture and CSF Analysis

This method is valuable in diagnosing viral infections of the CNS. CSF analysis typically involves determining the patients total white cell count, glucose level, and protein level in the CSF. Viral infection of the CNS tends to increase the total white cell count, while increasing the level of protein. The levels of glucose tend to be decreased by viral infection, due to an increased glucose consumption.

CSF Nucleic Acid Amplification using Polymerase Chain Reaction (PCR)

PCR is frequently used to for rapid identification of specific DNA viruses from the CSF, while Reverse transcriptase PCR is commonly used to identify RNA viruses in the CSF. The accuracy of this diagnostic tool is limited by the amount of the virus present in the CSF. Viral replication tends to peak early and then decline to undetectable levels in CNS infection. Within the first 5 days of symptom onset, before the decline of viral replication, PCR assays have a higher incidence of detecting CNS infection.

Serology

Serology is useful in diagnosing viral infections of the CNS when PCR analysis returns negative results.

Brain Biopsy

In recent years, due to the development of less invasive diagnosis techniques, brain biopsies are no longer frequently used for diagnosing viral infections of the nervous system. However, some viral infections of the CNS cannot be diagnosed without histological and electron microscopic evidence. In these cases, brain biopsies are only performed when the patient has a serious neurological illness and is in need of immediate therapy, an alternative procedure will not lead to a specific diagnosis, and the information gained by the brain biopsy will outweigh the risks.

Research and Therapy

Use of Antivirals to Treat CNS Infection

The use of antiviral treatment with both Multiple Sclerosis and AIDS dementia has proven ineffective as a treatment. In patients with Multiple Sclerosis, antiviral treatment of EBV with Acyclovir showed no significant difference from the placebo. In patients with AIDS dementia, despite antiretroviral therapy, CNS function remains diminished.

Use of Viruses for Gene Therapy

HSV-1 is a promising gene therapy agent, which could be used for gene delivery to neurons. This therapy may be used to treat metabolic brain diseases, neurodegenerative disorders, or to help enhance repair of brain tissue in neurological diseases.

Molecular Virology

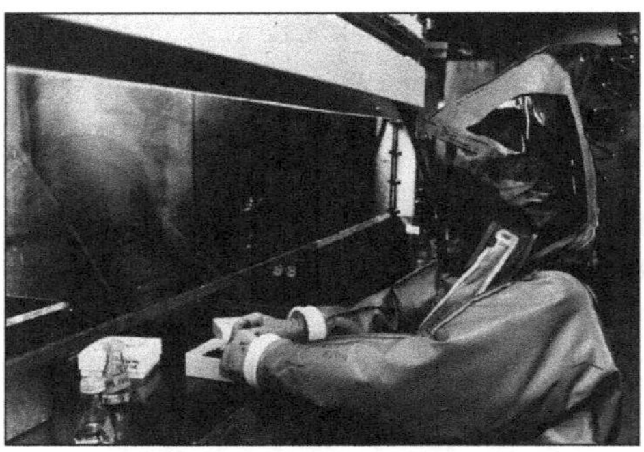

CDC worker in the maximum containment virology laboratory.

Molecular virology is the study of viruses on a molecular level. Viruses are submicroscopic parasites that replicate inside host cells. They are able to successfully infect and parasitize all kinds of life forms- from microorganisms to plants and animals- and as a result viruses have more biological diversity than the rest of the bacterial, plant, and animal kingdoms combined. Studying this diversity is the key to a better understanding of how viruses interact with their hosts, replicate inside them, and cause diseases.

Viral Replication

Viruses rely on their host to replicate and multiply. This is because viruses are unable to go through cell division, as they are acellular–meaning they lack the genetic information that encode the necessary tools for protein synthesis or generation of metabolic energy; hence they rely on their host to replicate and multiply. Using the host cell's machinery the virus generates copies of its genome and produces new viruses for the survival of its kind and the infection of new hosts. The viral replication process varies depending on the virus's genome.

Classification

In 1971 David Baltimore, a Nobel Prize-winning biologist, created a system called Baltimore Classification System.According to this system, viruses are classified into seven classes based on their replication strategy:

- Class I: Double-stranded DNA. Depending on the location of genome replication, this class can be subdivided into two groups: (a) viruses in which replication happens exclusively in the nucleus and is thus relatively dependent on cellular factors; (b) viruses replicating in cytoplasm that are mostly independent of cellular machinery due to the fact that they have acquired all needed factors for their genome's transcription and replication.
- Class II: Single-stranded DNA. These viruses only replicate in the nucleus. A double-stranded intermediate is formed during the replication process that serves as a template for the synthesis of virus's single-stranded DNA.
- Class III: Double-stranded RNA. The viruses in this class have segmented genomes. Each segment is transcribed individually to produce a monochromatic mRNA that codes for only one protein.
- Class IV: Single-stranded RNA - Positive-sense. The genome of these viruses are positive-sense RNAs, in which the RNA is directly translated into a viral protein. These viruses can be subdivided into two groups depending on their translation process: (a) viruses with polycistronic mRNA, in which the RNA translates into multiple protein products; (b) viruses with more complex transcription than the first group where either subgenomic RNAs or two rounds of translation are necessary to produce the genomic RNA.
- Class V: Single-stranded RNA - Negative-sense. All viruses in this class have negative-sense RNAs, in which the RNA complements mRNA, transcribes into a positive-sense RNA via a viral polymerase, and translates into viral protein. The viruses in this class either have segmented or unsegmented RNA; either way, replication occurs within the cell's cytoplasm.
- Class VI: Single-stranded positive-sense RNA with DNA intermediate. These viruses use reverse transcriptase to convert the positive-sense RNA (as the template) into DNA. Retroviruses are the most well-known family among this class.

- Class VII: Double-stranded DNA with RNA intermediate. The genome of these viruses are gapped, double-stranded, and subsequently become filled to form cccDNA (covalently closed circular DNA). This group also use the reverse transcription during the process of maturation.

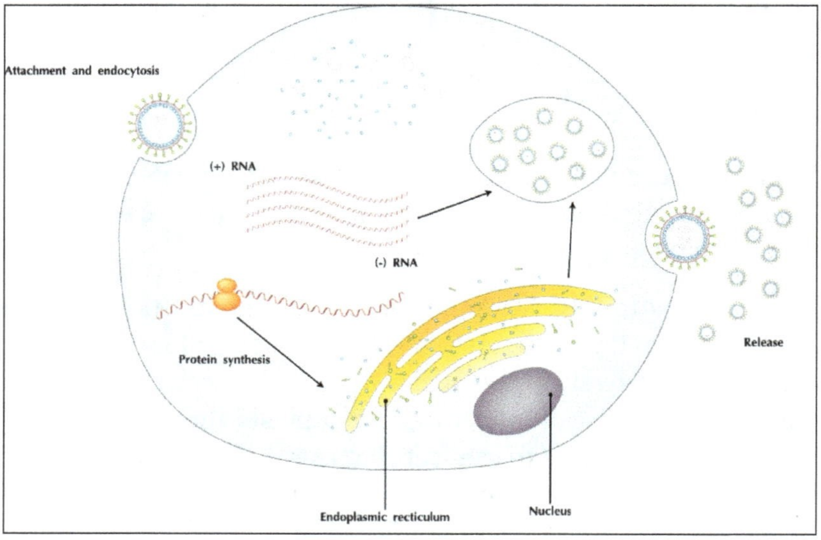

Viral replication cycle – simplified.

Replication Cycle

Regardless of the differences among viral species, they all share six basic replication stages:

Attachment is the cycle's starting point and consists of specific bindings between anti-receptors (or virus-attachment proteins) and cellular receptors molecules such as (glyco)proteins. The host range of a virus is determined by the specificity of the binding. The attachment causes the viral protein to change its configuration and thus fuse with the host's cell membrane; thereby enabling the virus to enter the cell.

Penetration of virus happens either through membrane fusion or receptor-mediated endocytosis and leads to viral entry. Due to their rigid cellulose-made (chitin in case of fungal cells) cell walls, plants and fungal cells get infected differently than animal cells. Often, a cell wall trauma is required for the virus to enter the cell.

Uncoating is the removal of viral capsid, which makes the viral nucleic acids available for transcription. The capsid could have been degraded by either host or viral enzymes, releasing the viral genome into the cell.

Replication is the multiplication of virus's genetic material. The process includes the transcription of mRNA, synthesis, and assembly of viral proteins and is regulated by protein expression.

Assembly process involves putting together and modifying newly-made viral nucleic acids and structural proteins to form the virus's nucleocapsid.

Release of viruses could be done by two different mechanisms depending on the type of virus. Lytic viruses burst the cell's membrane or wall through a process called lysis in order to release themselves. On the other hand, enveloped viruses become released by a process called budding in which a virus obtain its lipid membrane as it buds out of the cell through membrane or intracellular vesicle. Both lysis and budding processes are highly damaging to the cell, with the exception of retroviruses, and often result in the cell's death.

Viral Pathogenesis

Pathogenicity is the ability of one organism to cause disease in another. There is a specialized field of study in virology called viral pathogenesis in which it studies how viruses infect their hosts at the molecular and cellular level.

In order for the viral disease to develop several steps need to be taken. First, the virus has to enter the body and implant itself into a tissue (e.g. respiratory tissue). Second, the virus has to reproduce extracellular after invading in order to make ample copies of itself. Third, the synthesized viruses must spread throughout the body via circulatory systems or nerve cells.

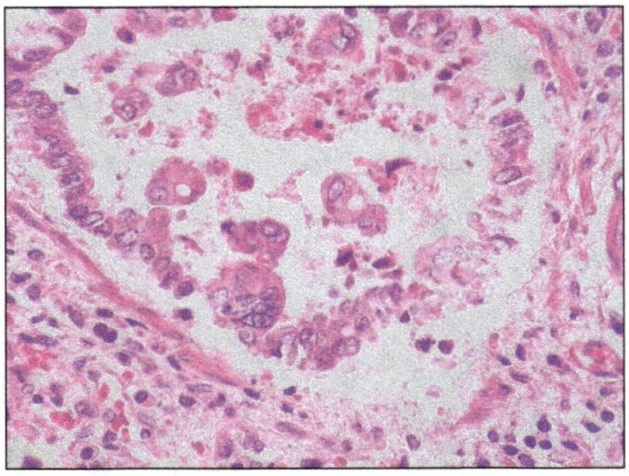

Respiratory syncytial virus (RSV) infection usually produces widespread bronchiolitis and interstitial pneumonia which may sometimes by associated with giant cells. This image shows a non-specific interstitial pneumonia pattern with no giant cells present.

With regards to viral diseases it is essential to look at two aspects: (a) the direct effect of virus replication; (b) the body's responses to infection. The course and severity of all viral infections are determined by the dynamic between the virus and the host. Common symptoms of virus infections are fever, body aches, inflammation, and skin rashes. Most of these symptoms are caused by our immune system's response to infection and are not the direct effect of viral replication.

Cytopathic Effects of Virus

Often it is possible to recognize virus-infected cells through a number of common phenotypic changes referred to as cytopathic effects. These changes include:

- Altered shape: The shape of adherent cells– cells that attach themselves to other cells or artificial substrates— may change from flat to round. In addition, the cell's extensions resembling tendrils – involved in cell's mobility and attachments—are withdrawn into the cell.
- Detachment from the substrate: Happens as a result of adherent cells' damage. The cell damage is due to partial degradation of cytoskeleton.
- Lysis: In this case the entire cell breaks down because of the absorption of extracellular fluid and swelling. Not all viruses cause lysis.
- Membrane fusion: The adjacent cells' membranes fuse together and produce a mass called syncytium, in which the cytoplasm contains more than one nucleus and appears as a giant cell. These cells have a short lifespan compared to other cells.
- Increase in membrane permeability: Viruses can increase the membrane permeability which allows many extracellular ions(e.g. iodine and sodium ions) to enter the cell.

Viral Infections

There are five different types of viral infections:

- Abortive infection: This kind of infection occurs when a virus successfully invades a host cell but is unable to complete its full replication cycle and produce more infectious viruses.

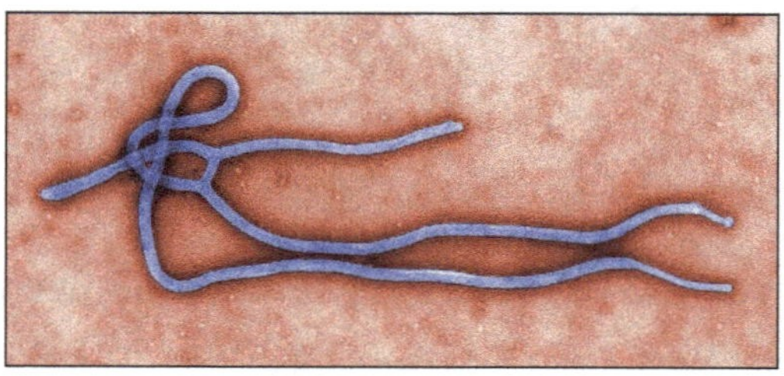

Ebola virus virion: This transmission electron micrograph (TEM) revealed some of the ultrastructural morphology displayed by an Ebola virus virion.

- Acute infection: Many common viral infections fallow this pattern. Acute infections are brief since they are often completely eliminated by the immune system. Acute infection is frequently associated with epidemics since most of virus replication happens before the onset of symptoms.
- Chronic infection: These infections have a prolonged course and are hard to eliminate since the virus stays in the host for a significant period.

- Persistent infection: There is a delicate balance between the host and the virus in this pattern. The virus adjusts its replication and pathogenecitiy levels to keep the host alive for its own benefit. While it is possible for the virus to live and replicate inside the host for its entire lifetime, oftentimes the host eventually eliminates the virus.
- Latent infection: Being the ultimate infection, the latent virus infections tend to exist inside the host for its entire lifetime. A well-known example of such infection is the herpes simplex virus in humans. This virus is able to stop its replication and restrict its gene expression in order to stop the recognition of the infected cell by the host's immune system.

Prevention and Treatment

Viral vaccines contain inactivated viruses which have lost their ability to replicate. These vaccines can prevent or lower the intensity of viral illness. Developing vaccines against smallpox, polio, and hepatitis B over the past 50 years has had a significant impact on world health and thus on global population. Nevertheless, there have been ongoing viral outbreaks (such as the Ebola and Zika viruses) in the past few years affecting millions of people all around the globe. Therefore, more advanced understanding of molecular virology and viruses are needed for the development of new vaccines and the control of ongoing/future viral outbreaks. In Europe, several countries put a higher focus on Virology research compared to other member states. These countries include Estonia, Finland, Denmark and Latvia.

An alternative way to treat viral infections would be antiviral drugs in which the drug blocks the virus's replication cycle. The specificity of an antiviral drug is the key to its success. These drugs are toxic to both the virus and the host but in order to minimize their damage they are developed in such a way as to be more toxic to the virus than to the host. The efficiency of an antiviral drug is measured by the chemotherapeutic index, given by:

$$\frac{\text{Dose of drug that inhibits virus replication}}{\text{Dose of drug that is toxic to host}}$$

The higher the drug's efficiency, the smaller the value of chemotherapeutic index. In clinical practice, this index is used to produce a safe and clinically useful drug.

Veterinary Virology

Veterinary virology is the study of viruses in non-human animals. It is an important branch of veterinary medicine.

Rhabdoviruses

Rhabdoviruses are a diverse family of single stranded, negative sense RNA viruses that infect a wide range of hosts, from plants and insects, to fish and mammals. The Rhaboviridae family consists of six genera, two of which, cytorhabdoviruses and nucleorhabdoviruses, only infect plants. Novirhabdoviruses infect fish, and vesiculovirus, lyssavirus and ephemerovirus infect mammals, fish and invertebrates. The family includes pathogens such as rabies virus, vesicular stomatitis virus and potato yellow dwarf virus that are of public health, veterinary, and agricultural significance.

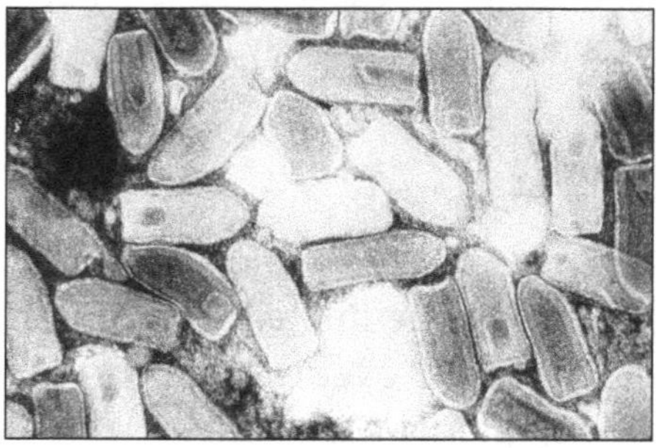

Vesicular stomatitis virus (VSV), the prototypical rhabdovirus.

Pestiviruses

Pestiviruses have a single stranded, positive-sense RNA genomes. They cause Classical swine fever (CSF) and Bovine viral diarrhea(BVD). Mucosal disease is a distinct, chronic persistent infection, whereas BVD is an acute infection.

Arteriviruses

Arteriviruses are small, enveloped, animal viruses with an icosahedral core containing a positive-sense RNA genome. The family includes equine arteritis virus (EAV), porcine reproductive and respiratory syndrome virus (PRRSV), lactate dehydrogenase elevating virus (LDV) of mice and simian haemorrhagic fever virus (SHFV).

Coronaviruses

Coronaviruses are enveloped viruses with a positive-sense RNA genome and with a nucleocapsid of helical symmetry. They infect the upper respiratory and gastrointestinal tract of mammals and birds. They are the cause of a wide range of diseases in cats, dog, pigs, rodents, cattle and humans. Transmission is by the faecal-oral route.

Toroviruses

Torovirus is a genus of viruses within the family Coronaviridae, subfamily Torovirinae that primarily infect vertebrates and include Berne virus of horses and Breda virus of cattle. They cause gastroenteritis in mammals, including humans but rarely.

Influenza

Influenza is caused by RNA viruses of the family Orthomyxoviridae and affects birds and mammals.

Avian Influenza

Wild aquatic birds are the natural hosts for a large variety of influenza A viruses. Occasionally viruses are transmitted from this reservoir to other species and may then cause devastating outbreaks in domestic poultry or give rise to human influenza pandemics.

Swine Influenza

Swine influenza is an infection caused by any one of several types of swine influenza viruses. Swine influenza virus (SIV) or swine-origin influenza virus (S-OIV) is any strain of the influenza family of viruses that is endemic in pigs. As of 2009, the known SIV strains include influenza C and the subtypes of influenza A known as H1N1, H1N2, H2N1, H3N1, H3N2, and H2N3.

The swine flu was initially seen in Mexico in April 2009, where the strain of the particular virus was a mixture from 3 types of strains. Six of the genes are very similar to the H1N2 influenza virus that was found in pigs around 2000.

Bluetongue Virus

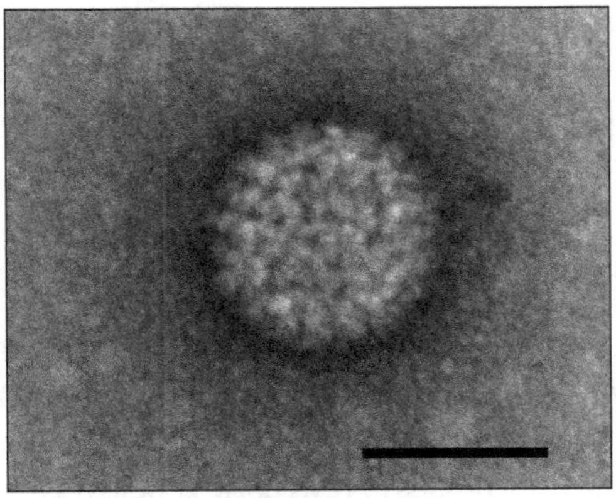

Bluetongue virus.

Bluetongue virus (BTV), a member of Orbivirus genus within the Reoviridae family causes serious disease in livestock (sheep, goat, cattle). It is non-enveloped, double-stranded RNA virus. The genome is segmented.

Circoviruses

Circoviruses are small single-stranded DNA viruses. There are to genera: gyrovirus, with one species called chicken anemia virus; and circovirus, which includes porcine circovirus types 1 and 2, psittacine beak and feather disease virus, pigeon circovirus, canary circovirus, goose circovirus.

Herpesviruses

Herpesviruses are ubiquitous pathogens infecting a variety of animals, including humans. Hosts include many economically important species such as abalone, oysters, salmon, poultry (avian infectious laryngotracheitis, Marek's disease), cattle (bovine malignant catarrhal fever), dogs, goats, horses, cats (feline viral rhinotracheitis), and pigs (pseudorabies). Infections may be severe and may result fatality, so outbreaks of herpesviruses in livestock cause significant financial losses, and are thus an important area of study in veterinary virology.

Retroviruses

Retroviruses are established pathogens of veterinary importance. They are generally a cause of cancer or immune deficiency.

Flaviviruses

Flaviviruses constitute a family of linear, single-stranded RNA(+) viruses. Flaviviruses include the West Nile virus, dengue virus, Tick-borne Encephalitis Virus, Yellow Fever Virus, and several other viruses. Many flavivirus species can replicate in both mammalian and insect cells. Most flaviviruses are arthropod borne and multiply in both vertebrates and arthropods. The viruses in this family that are of veterinary importance include Japanese encephalitis virus, St. Louis encephalitis virus, West Nile virus, Israel turkey meningoencephalomyelitis virus, Sitiawan virus, Wesselsbron virus, yellow fever virus and the tick-borne flaviviruses e.g. louping ill virus.

Paramyxoviruses

Paramyxoviruses are a diverse family of non-segmented negative strand RNA viruses that include many highly pathogenic viruses affecting humans, animals, and birds. These include canine distemper virus (dogs), phocine distemper virus (seals), cetacean morbillivirus (dolphins and porpoises) Newcastle disease virus (birds) and rinderpest virus (cattle). Some paramyxoviruses such as the henipaviruses are zoonotic pathogens, occurring primarility in an animal hosts, but also able to infect humans.

Parvovirus

Parvoviruses are linear, non-segmented single-stranded DNA viruses, with an average genome size of 5000 nucleotides. They are classified as group II viruses in Baltimore classification of viruses. Parvoviruses are among the smallest viruses and are 18–28 nm in diameter.

Parvoviruses can cause disease in some animals, including starfish and humans. Because the viruses require actively dividing cells to replicate, the type of tissue infected varies with the age of the animal. The gastrointestinal tract and lymphatic system can be affected at any age, leading to vomiting, diarrhea and immunosuppression but cerebellar hypoplasia is only seen in cats that were infected in the womb or at less than two weeks of age, and disease of the myocardium is seen in puppies infected between the ages of three and eight weeks.

Paleovirology

"Paleovirology" is the study of ancient extinct viruses (called "paleoviruses") and the effects that these agents have had on the evolution of their hosts. Thus far, the study of these viruses has mostly been limited to endogenous retroviruses that can be directly identified from their remnants in host genomes. However, one can infer the existence of other paleoviruses from their evolutionary pressures on host genes. We suggest that selection to survive the pathogenic effects of these viruses has shaped our repertoire of antiviral defenses in ways that impact our resistance or susceptibility to modern-day emerging viruses.

Unearthing Signs of Ancient Viral Infections

The human genome is a living document of ancient and now extinct viruses. Indeed, DNA of retroviral origin makes up 8% of human genome sequence. Retroviruses are RNA viruses that replicate through a DNA intermediate called a provirus, which becomes integrated into the host cell chromosome. Usually such integrations occur in somatic cells, but when integration of the provirus occurs in a germ cell, an endogenous retrovirus can be inherited as part of the genome. If these germline insertions become fixed within a population, the provirus becomes part of the genetic legacy of the species. It is difficult to calculate exactly how many retroviral infections of the germline led to the ~100,000 copies of endogenous retroviruses in the human genome because duplications, transpositions, and other non-infectious events also contribute to this number. However, at a very minimum, each of the more than 31 families of endogenous retrovirus found in the human genome must have arisen from one or more separate paleoviruses that infected the ancestors of modern humans. Since reinfections of the germline with members of the same families occurred frequently, retroviral infections that impacted the genome had to have happened repeatedly during primate evolution with the most recent episode in humans between 100,000 and 1 million years ago.

The dashed line at the top line represents the time period for paleoviruses. The red lightning bolts represent dates of known recent and ancient viruses based either on historical records, molecular clocks, or on endogenous retroviruses in the human genome for SARS-Co, HIV-1 and -2, dengue (DEN), measles, smallpox, HERV-K(HML2), or PtERV, and older endogenous retroviruses shared among all hominoids or all primates. The blue, green, and orange lightning bolts represent inferred viruses based on positive selection of TRIM5, and the brown line is an inferred virus based on positive selection of ZAP simplified for representation, here. Each color corresponds to inferred paleoviruses based on positive selection on a particular antiviral gene calculated on the phylogenetic tree in the bottom of the figure where the lineage under selection has the same color coding, and the dates correspond to dates of the ancestors. Although one virus is shown per episode of selection, there could be many different waves of similar viruses during that time period. Purple branches refer to selections due to inferred paleoviruses in lineages that do not lead directly to humans. There is considerable uncertainty associated with most of the dates referred to in this figure.

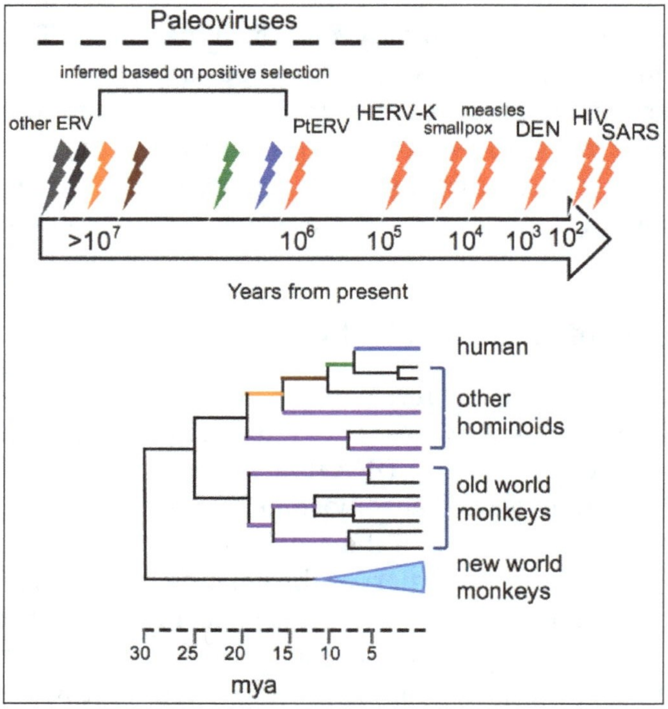

Time-line of paleoviruses in the human lineage.

This impressive fossil record, represented by endogenous retroviruses, is still likely a vast underestimate of the number of retroviral infections of human ancestors. Many retroviruses do not infect the germline. For example, the human T cell leukemia virus types I and II (HTLVI and II) are thought to have entered human populations over 20,000 years ago, but no endogenous copies in human genomes have yet been found. For those retroviruses that do infect the germline, the majority of integration events did not become fixed since they were negatively selected out of the population or lost by drift. Indeed, the integration pattern of endogenous retroviruses (usually located away from genes) is quite different from the integration pattern of their exogenous counterparts (often located near or in genes), suggesting that selection has cleansed the majority of endogenization events from primate genomes. These imperfections in the fossil record of endogenous retroviruses can lead to vast underestimation of the age of viral lineages, as was recently demonstrated for the lentiviruses (retroviruses such as HIV-1) where the finding of endogenized copies in two independent lemur genera upwardly revised the age of primate lentiviruses to more than 4 million years.

It has been challenging for paleovirology to move beyond the study of endogenous retroviruses since it is much more difficult to decipher ancient viruses that left no extant copies of their past existence. The recent finding that at least one bornavirus gene has integrated in several mammalian genomes at multiple evolutionary periods demonstrates both the possibility of identifying and dating some other ancient classes of viral infections although this remains a rare event outside of the retrovirus family. Nonetheless, endogenous copies of viruses are but a subset of the ancient and extinct viruses that we call paleoviruses. The existence of some additional paleoviruses can be deciphered by estimating the age of the last common ancestor of extant viruses. For viruses where there is evidence for co-speciation of virus and host, we can arrive at a conservative estimate of their evolutionary age by correlation to the divergence of their hosts. The best example of primate viruses that have cospeciated with its host over a considerable evolutionary period are the spumaretroviruses. For other viruses, like herpesviruses and papillomaviruses, there is also some support for host–virus cospeciation.

In contrast, DNA viruses such as orthopoxviruses show ample evidence of host switching and have no suitable cospeciation estimates. In such cases, although we can calculate the time to the last common ancestor from estimates of mutation rates in extant sequences, this is likely to be an underestimate due to the extinction of viral lineages (and thus, the loss of their sequence diversity in the calculation). Moreover, these age estimates do not provide any information as to which host lineages might have been affected by any given inferred paleovirus. However, tracing the acquisitions of genes and retroelements from host genomes into large DNA viruses may provide an additional means to estimate their divergence times. On the other hand, the case for ancient RNA viruses, which provide more examples of pathogenic viruses than DNA viruses, is even more vexing because high mutation rate essentially obliterates bioinformatic signals and our ability to meaningfully estimate ancestry beyond a few million years. Indeed, based on mutation rates and sampled lineages, one might incorrectly infer that the last common ancestor for most RNA viruses was very young.

Evolutionary Signal of Ancient Infections within Antiviral Genes

Given these problems in using the sequences of current day viruses to identify most paleoviruses, we propose another approach to infer the existence of pathogenic paleoviruses. This approach is based on a striking evolutionary signature called positive selection that these viruses impart on antiviral genes in primate genomes. When a viral infection rages in a population, pre-existing variants of these antiviral defense genes are acted upon by Darwinian selection, leading to the accelerated fixation of even previously rare variants in the species. It is important to emphasize that the inability to ward off a viral infection can translate to significant fitness costs. Therefore, beneficial variants in antiviral genes will spread to fixation by virtue of their significant selective advantage even if viral infection does not lead to death. Counterevolution by the virus (or introduction of new viruses that evade the host defenses using the same mechanisms) eventually leads to renewed selective pressure on antiviral genes. These recurrent bouts of selection represent evolutionary arms races that can be detected by comparing DNA coding sequences of related species and by using maximum likelihood methods looking for excesses of mutations that change amino acids in gene sequences (non-synonymous mutations) relative to mutations that do not change amino acids (synonymous mutations). Repeated episodes of these arms races between host antiviral genes and new viral challenges will lead to dramatic rates of change of non-synonymous mutations. Thus, signatures of evolutionary changes in protein-coding sequences of antiviral genes allow us to infer the selection of ancient host species due to paleoviruses.

Of course, it would be incorrect to imply that the complete outcome of a viral infection is decided on the basis of a single host antiviral gene. If this were true, then the differences in tempo of viral versus host evolution would make it unlikely that the hosts could ever adapt away from a viral infection. Instead, the host–virus interaction is more complex, where alleles in single host genes vary in relative resistance/susceptibility to viruses. The outcome of a particular viral infection on a population level is determined by the compendium of antiviral genes borne by the host, the viral fitness cost associated with escape from antiviral genes, as well as many factors beyond host and viral genetics. Nonetheless, over evolutionary timescales, the effects of single genes can be discerned as long as beneficial alleles confer fitness advantages to the host.

The comparison of sequence data from multiple orthologs of primate antiviral genes allows us to not only infer that a given gene was under positive selection, but also when it was under positive selection. This can be done by reconstructing the ancestral coding sequence of an antiviral gene at each node of a primate phylogenetic tree in silico and then determining where positive selection occurred on internal nodes (the internal nodes can be dated by other molecular evolutionary methods, assuming some sort of molecular clock for mutation rates in primate genomes, or by paleontological fossil records). Each case of positive selection for an antiviral gene on the phylogenetic tree then implies the presence of at least one paleovirus on that branch. For instance, since diverging from the chimpanzee lineage, human TRIM5, a potent antiviral factor, has incurred eight non-synonymous and two synonymous changes—this roughly translates to an excess of three non-synonymous changes over what might be expected by chance. This information suggests the presence of at least one paleovirus that was in genetic conflict with TRIM5 in the 4–5 million years in the human lineage since the human–chimpanzee divergence. However, there is no information about when each of these three changes became fixed in the population—this could have occurred in rapid succession or slowly over 4–5 million years. When the positive selection of different unlinked antiviral genes are compared to one another, one finds different patterns of episodic positive selection in primate evolution for each gene. This suggests that different ancient viral pathogens acting at different times were responsible for driving selection of one gene versus another. In theory, episodes of positive selection could also occur on multiple antiviral genes due to the same pathogenic virus if the fitness costs imposed by a particular virus were especially severe at any given period of primate evolution.

Another means to date a paleoviral infection is the finding of a gene fusion event in a particular primate lineage that resulted in novel antiviral activity, exemplified by the TRIM5 fusion to a CypA retrogene in the Aotus genus of New World monkeys. Dating the origin of such a gene fusion can be informative even if the "newly born" antiviral gene eventually degrades due to relaxed selection (once the paleoviral challenge is extinguished). In both cases (positive selection or gene fusions), the time-period under selection can only be ascribed to a branch on the evolutionary tree, which may represent several million years of evolution (the age represented by the branch is determined by the phylogenetic density of extant species and extent of sequence sampling). Other methods can examine the positive selection on antiviral genes in shorter time-scales by looking for evidence of selective sweeps using single nucleotide polymorphisms in modern population data, for instance in human populations. These methods are especially useful in determining the action of a relatively "modern" selective pathogen which may or may not also have had an ancient counterpart. Finding both modern and ancient selective episodes is strong indication of a particular antiviral gene having been repeatedly acted upon by antagonism against both ancient paleoviruses and present-day viral pathogens.

Challenges and Opportunities in using Evolution to Infer Paleovirology

There are limitations to inferring episodes of paleoviral infection from signatures of positive selection. Some weaknesses in the method arise directly from methodology; by definition, it is difficult, if not impossible, to detect a single episode of positive selection if it occurred in the distant past. Thus, for the most part, we are left considering only those antiviral genes that have been repeated targets of selection. Moreover, our ability to use reconstructive evolution to infer positive selection may weaken at deep evolutionary distances, although the 35-million-year age of simian primate evolution is especially tractable. An additional limitation is the fact that the same antiviral gene may have acted on different viruses over the evolutionary record, with each new episode potentially "over-writing" the previous one. Each instance of adaptation may involve only a few amino acid changes, but if these occur at the same position then the record can get muddled in terms of elucidating exactly which change was selected when. This limitation is exemplified by the human major histocompatibility (MHC) class I genes devoted to the presentation of viral antigens to the adaptive immune system. The vast diversity and rapid evolution of alleles in MHC Class I demonstrates that humans and primate ancestors have undergone dramatic episodes of positive selection to protect against continual and diverse viral attack; however the constant turnover has eroded information about all but the most recently encountered viruses. Finally, the evolutionary analysis of antiviral genes that also play a role in limiting other microbial pathogens would confound the assessment of what type of pathogen was responsible for the positive selection.

Notwithstanding these caveats, we believe it is possible to identify candidate classes of paleoviruses based on analyses of positive selection in antiviral genes. First, our knowledge of the functions of antiviral genes under selection provides clues to which viruses might be considered. For instance, some antiviral genes are only active against a limited range of viral lineages. Thus, ancient episodes of positive selection in TRIM5 are likely indicative of ancient retroviral infections where the evolutionary loss of the ability of TRIM5 to combat HIV shows that ancient selective events that changed the specificity of TRIM5 clearly have had a detrimental effect on the resistance of modern humans to this contemporary virus. Moreover, even when the antiviral gene is embattled with several viruses simultaneously, the amino acid positions that have evolved under positive selection can be a good indication of what kind of antagonism was encountered during a given evolutionary period. For example, even through protein kinase R can act as a general antiviral protein, evolutionary reconstructions have allowed us to infer the presence of virus-encoded mimics of eIF2alpha throughout the history of primate evolution, despite the fact that such mimics are known in only a few extant and relatively young lineages of poxviruses and iridoviruses.

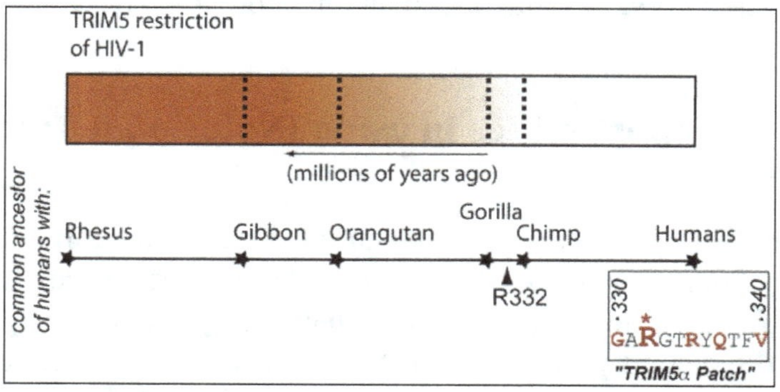

TRIM5 restriction of HIV-1 has decreased during evolution leading to humans.

The shading of the rectangle represents the degree that TRIM5 will limit infection of HIV-1 (darker color means TRIM5 decreases HIV-1 infection more) and the X-axis indicates time in millions of years from the present. Each dotted line represents the reconstruction of TRIM5 as it likely existed at a node of a phylogenetic tree indicating a common ancestor of humans with chimpanzees, gorillas, orangutans, gibbons, and Old World monkeys (rhesus). Original data is found in and shows that the antiviral gene TRIM5 restricted HIV-1 better at points in evolution earlier than the chimp–human common ancestor than it does after that. On the right shows an amino acid sequence of a region of TRIM5 containing amino acids that confer resistance or susceptibility to HIV-1 with the amino acids that are under the strongest positive selection in red. Changes in this region cause a gain of restriction to some viruses, while causing a loss to others. The R332 amino acid, which represents the single largest determinant of loss of resistance to HIV-1, was fixed before the chimp–human common ancestor, but positive selection has continued in TRIM5 along the human lineage beyond this point.

Second, since the rapid evolution of host antiviral genes is likely driven by their genetic conflict with specific viral antagonists, the classification of which modern viruses can neutralize the function of which antiviral genes under positive selection can help define classes of possible candidate ancient viral antagonists and thereby pathogens. Similarly, there may be cases of "missing" viral species where extinction of a virus that is present in sister primate species could indicate the aftermath of a virus–host interaction that was beneficial to one host lineage. Examples include the PtERV retrovirus that is not present in human genomes but is present in chimpanzee and gorilla genomes, a lineage of rhadinovirus that has not yet been found in humans but is present in sister taxa, and foamy viruses that are present in most primates but not humans. Finding a human gene active against a virus that is not found in humans could be a starting point for considering whether or not an older version of that virus was eliminated from the evolutionary lineage leading to modern humans due to genetic adaptation rather than geographical isolation.

Purpose of Paleovirology

Paleovirology could be viewed as the study of ancient viruses that primate genomes encountered and defeated during the course of evolution. This view emphasizes that our current antiviral repertoire was not optimized to combat present infections, but rather is the product of selection for survival of our species' past infections. Thus, the selective changes that these antiviral genes incurred during these periods of evolutionary pressure might make them less competent to fight modern viral challenges. For example, the human TRIM5 gene does not inhibit HIV, although it was certainly selected to inhibit something else in our past. The analysis of amino acids on antiviral genes driven by selection of ancient pathogens can be used to identify the interface between the host protein and the virus in ways that could conceivably be used to design rational antiviral drugs or gene therapy strategies. Such analyses of the virus–host battles, on an evolutionary scale, can also explain the otherwise mysterious loss of antiviral activities. For instance, antiviral genes that serve no other cellular functions can incur significant fitness costs or relaxed selection and therefore can be lost due to the lack of a pathogen during extended periods of time. These changes and losses to our antiviral repertoire may help explain deficiencies in our current innate immune protection against pathogenic viral challenges.

The study of "resurrected" paleoviruses, exemplified by the evolution-guided reconstruction of several endogenous retroviruses and the 1918 influenza virus can also reveal previously hidden details of host–virus interactions. Finally, the universe of possible "emerging viruses" that could arise by cross-species transmission of viruses from other animals into humans is still largely undefined. It is possible that amidst this reservoir of potential pathogens lie the descendants of viruses that were once eliminated from human ancestors, but because of continued evolution of both virus and humans could now reinfect modern humans. Although we have focused on paleoviruses of humans and primate ancestors in this essay, the same arguments can be made and lines of research can be applied to nearly any set of animal or plant families for which virus–host relationships exist. Paleovirology, in this broader sense, may be able to correlate the existence of ancient infections with known phylogeographical events, such as extinctions, bursts of speciation, exchanges of fauna between continents, island isolations, and population migrations.

References

- Clinical-virology, topics-medicine-and-dentistry: sciencedirect.com, Retrieved 14 June, 2020
- Stauft, Charles B; Wimmer, Eckard (2018-02-16), "Synthetic Viruses", eLS, John Wiley & Sons, Ltd, pp. 1–7, doi:10.1002/9780470015902.a0027771, ISBN 9780470015902
- Dimmock, Nigel (2007). Introduction to Modern Virology. Wiley-Blackwell. ISBN 978-1-119-97810-7
- Cann, Alan (2012). Principles of Molecular Virology. ELSEVIER. p. 214. ISBN 9780123849403
- Edward J Dubovi & MacLachlan NJ (2010). Fenner's Veterinary Virology, Fourth Edition. Boston: Academic Press. p. 395. ISBN 978-0-12-375158-4

Classification of Viruses

3

Viruses can be categorized depending on the host cell they interact with. It is classified into plant viruses, animal viruses, fungal viruses and bacteriophages. The topics elaborated in this chapter will help in gaining a better perspective of the classification of viruses.

Viruses are classified on the basis of morphology, chemical composition, and mode of replication. The viruses that infect humans are currently grouped into 21 families, reflecting only a small part of the spectrum of the multitude of different viruses whose host ranges extend from vertebrates to protozoa and from plants and fungi to bacteria.

Morphology

Helical Symmetry

In the replication of viruses with helical symmetry, identical protein subunits (protomers) self-assemble into a helical array surrounding the nucleic acid, which follows a similar spiral path. Such nucleocapsids form rigid, highly elongated rods or flexible filaments; in either case, details of the capsid structure are often discernible by electron microscopy. In addition to classification as flexible or rigid and as naked or enveloped, helical nucleocapsids are characterized by length, width, pitch of the helix, and number of protomers per helical turn. The most extensively studied helical virus is tobacco mosaic virus. Many important structural features of this plant virus have been detected by x-ray diffraction studies. Figure shows Sendai virus, an enveloped virus with helical nucleocapsid symmetry, a member of the paramyxovirus family.

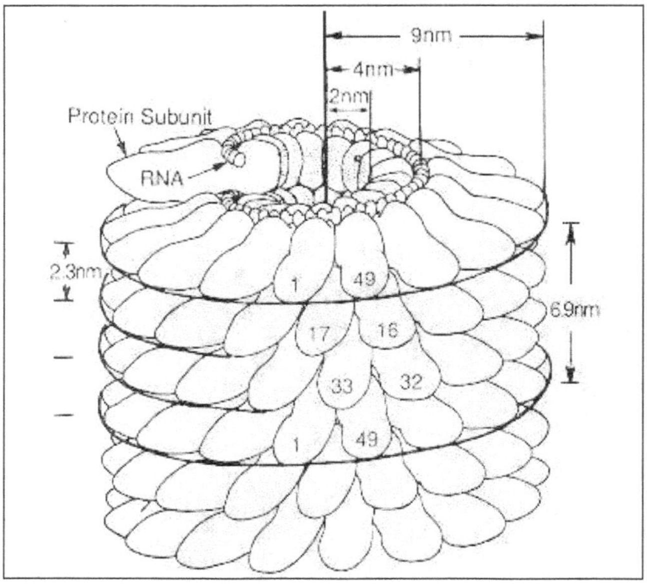

The helical structure of the rigid tobacco mosaic virus rod. About 5 percent of the length of the virion is depicted. Individual 17,400-Da protein subunits (protomers) assemble in a helix with an axial repeat of 6.9 nm (49 subunits per three turns). Each turn contains a nonintegral number of subunits (16-1/3), producing a pitch of 2.3 nm. The RNA (2×10^6 Da) is sandwiched internally between adjacent turns of capsid protein, forming a RNA helix of the same pitch, 8 nm in diameter, that extends the length of virus, with three nucleotide bases in contact with each subunit. Some 2,130 protomers per virion cover and protect the RNA. The complete virus is 300 nm long and 18 nm in diameter with a hollow cylindrical core 4 nm in diameter.

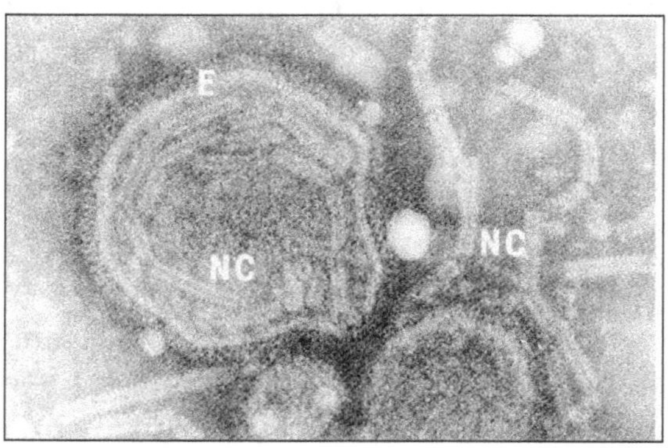

Fragments of flexible helical nucleocapsids (NC) of Sendai virus, a paramyxovirus, are seen either within the protective envelope (E) or free, after rupture of the envelope. The intact nucleocapsid is about 1,000 nm long and 17 nm in diameter; its pitch.

Icosahedral Symmetry

An icosahedron is a polyhedron having 20 equilateral triangular faces and 12 vertices. Lines through opposite vertices define axes of fivefold rotational symmetry: all structural features of the polyhedron repeat five times within each 360° of rotation about any of the fivefold axes. Lines through the centers of opposite triangular faces form axes of threefold rotational symmetry; twofold rotational symmetry axes are formed by lines through midpoints of opposite edges. An icosahedron (polyhedral or spherical) with fivefold, threefold, and twofold axes of rotational symmetryis defined as having 532 symmetry.

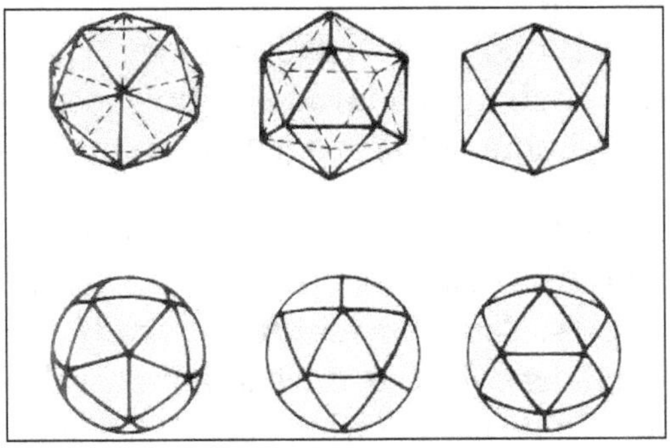

Icosahedral models seen, left to right, on fivefold, threefold, and twofold axes of rotational symmetry. These axes are perpendicular to the plane of the page and pass through the centers of each figure. Both polyhedral (upper) and spherical (lower) forms are represented by different virus families.

Viruses were first found to have 532 symmetry by x-ray diffraction studies and subsequently by electron microscopy with negative-staining techniques. In most icosahedral viruses, the protomers, i.e. the structural polypeptide chains, are arranged in oligomeric clusters called capsomeres, which are readily delineated by negative staining electron microscopy and form the closed capsid shell. The arrangement of capsomeres into an icosahedral shell permits the classification of such viruses by capsomere number and pattern. This requires the identification of the nearest pair of vertex capsomeres (called penton: those through which the fivefold symmetry axes pass) and the distribution of capsomeres between them.

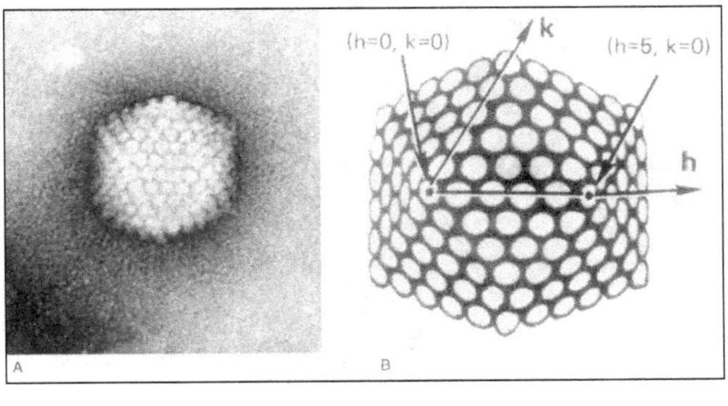

(A) The capsid reveals the typical isometric shell made up from 20 equilateral triangular faces. The 252 capsomeres, 12 pentons and the 240 hollow hexon capsomeres are arranged in a T = 25 symmetry pattern vetite.

(B) Adenovirus model: Capsomeres are depicted as circles surrounded by an electron dense stain. The inclined axes, h and k, are indicated. The second vertex has indices h = 5, k = 0. The total number of capsomeres $C = 10(h^2 + hk + k^2) + 2 = 252$. Capsomere organization is also expressed by the triangulation number, T, the number of unit triangles on each of the 20 faces of the icosahedron. A unit triangle is formed by lines joining the centers of three adjacent capsomeres. $T = (h^2 + hk + k^2) = 25$ for adenoviruses, and $C = 10T + 2$. The 12 vertex capsomeres are surrounded by 5 other capsomeres each, therefore called penton and show 5-fold rotational symmetry. The penton base consists of 5 identical 85 kD polypeptide chains and extrudes a long antenna-like fiber protein. The 240 hexon capsomeres are trimers of the 120 kD hexon protomere polypeptide

In the adenovirus model in left figure, one of the penton capsomeres is arbitrarily assigned the indices h = 0, k = 0 (origin), where h and k are the indicated axes of the inclined (60°) net of capsomeres. The net axes are formed by lines of the closest-packed neighboring capsomeres. In adenoviruses, the h and k axes also coincide with the edges of the triangular faces. Any second neighboring vertex capsomere has indices h = 5, k = 0 (or h = 0, k = 5). The capsomere number (C) can be determined to be 252 from the h and k indices and the equation: $C = 10(h^2 + hk + k^2) + 2$. This symmetry and number of capsomeres is typical of all members of the adenovirus family.

Virus Core Structure

Except in helical nucleocapsids, little is known about the packaging or organization of the viral genome within the core. Small virions are simple nucleocapsids containing 1 to 2 protein species. The larger viruses contain in a core the nucleic acid genome complexed with basic protein(s) and protected by a single- or double layered capsid (consisting of more than one species of protein) or by an envelope.

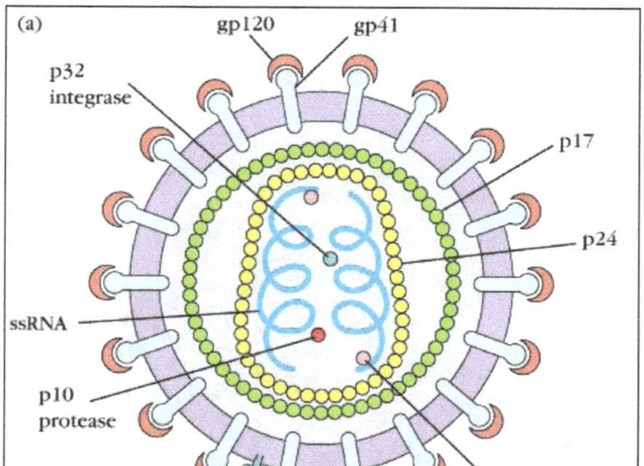

Two-dimensional diagram of HIV-1 correlating (immuno-) electron microscopic findings with the recent nomenclature for the structural components in a 2-letter code and with the molecular weights of the virus structural (glyco-) proteins. SU

stands for outer surface glycoprotein, TM for transmembrane gp, MA for membrane associated or matrix protein, LI for core-envelope-link, CA for major capsid, NC for nucleocapsid protein, respectively. PR, RT and IN represent the virus-coded enzymes protease, reverse transcriptase and integrase that are functional during the life cycle of a retrovirus

Chemical Composition and Mode of Replication

RNA Virus Genomes

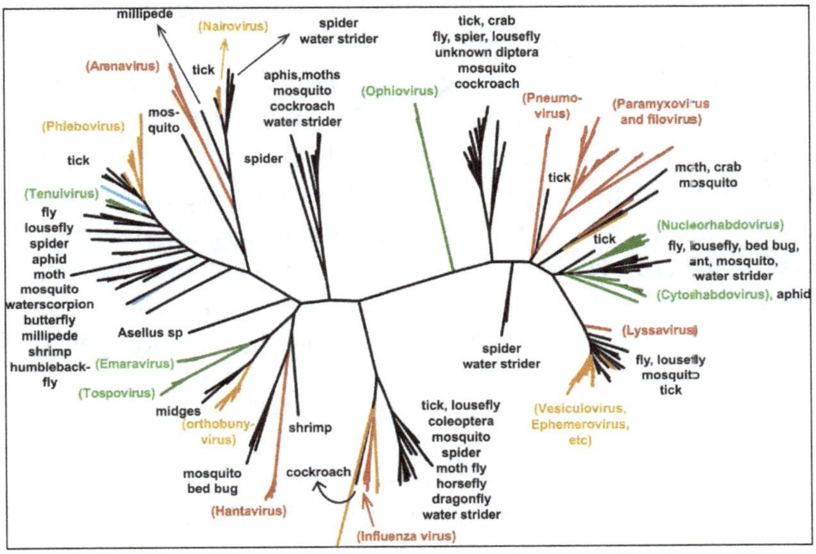

Schemes of 21 virus families infecting humans showing a number of distinctive criteria: presence of an envelope or (double-) capsid and internal nucleic acid genome. +, Sense strand; -, antisense strand; ±, dsRNA or DNA; o, circular DNA; C, number of capsomeres or holes, where known; nm, dimensions of capsid, or envelope when present; the hexagon designates the presence of an isometric or icosahedral outline.

RNA viruses, comprising 70% of all viruses, vary remarkably in genome structure. Because of the error rate of the enzymes involved in RNA replication, these viruses usually show much higher mutation rates than do the DNA viruses. Mutation rates of 10-4 lead to the continuous generation of virus variants which show great adaptability to new hosts. The viral RNA may be single-stranded (ss) or double-stranded (ds), and the genome may occupy a single RNA segment or be distributed on two or more separate segments (segmented genomes). In addition, the RNA strand of a single-stranded genome may be either a sense strand (plus strand), which can function as messenger RNA (mRNA), or an antisense strand (minus strand), which is complementary to the sense strand and cannot function as mRNA protein translation. Sense viral RNA alone can replicate if injected into cells, since it can function as mRNA and initiate translation of virus-encoded proteins. Antisense RNA, on the other hand, has no translational function and cannot per se produce viral components.

DsRNA viruses, e.g., members of the reovirus family, contain 10, 11 or 12 separate genome segments coding for 3 enzymes involved in RNA replication, 3 major capsid proteins and a number of smaller structural proteins. Each segment consists of a complementary sense and antisense strand that is hydrogen bonded into a linear ds molecule. The replication of these viruses is complex; only the sense RNA strands are released from the infecting virion to initiate replication.

The retrovirus genome comprises two identical, plus-sense ssRNA molecules, each monomer 7–11 kb in size, that are noncovalently linked over a short terminal region. Retroviruses contain 2 envelope proteins encoded by the env-gene, 4–6 nonglycosylated core proteins and 3 non-structural functional proteins (reverse transcriptase, integrase, protease: RT, IN, PR) specified by the gag-gene. The RT transcribes the viral ssRNA into double-stranded, circular proviral DNA. This DNA, mediated by the viral integrase, becomes covalently bonded into the DNA of the host cell to make possible the subsequent transcription of the sense strands that eventually give rise to retrovirus progeny. After assembly and budding, retroviruses show structural and functional maturation. In immature virions the structural proteins of the core are present as a large precursor protein shell. After proteolytic processing by the viral protease the proteins of the mature virion are rearranged and form the dense isometric or cone-shaped core typical of the mature virion, and the particle becomes infectious.

DNA Virus Genomes

Most DNA viruses contain a single genome of linear dsDNA. The papovaviruses, comprising the polyoma- and papillomaviruses, however, have circular DNA genomes, about 5.1 and 7.8 kb pairs in size. DsDNA serves as a template both for mRNA and for self-transcription. Three or 2 structural proteins make up the papovavirus capsid: in addition, 5-6 nonstructural proteins are encoded that are functional in virus transcription, DNA replication and cell transformation.

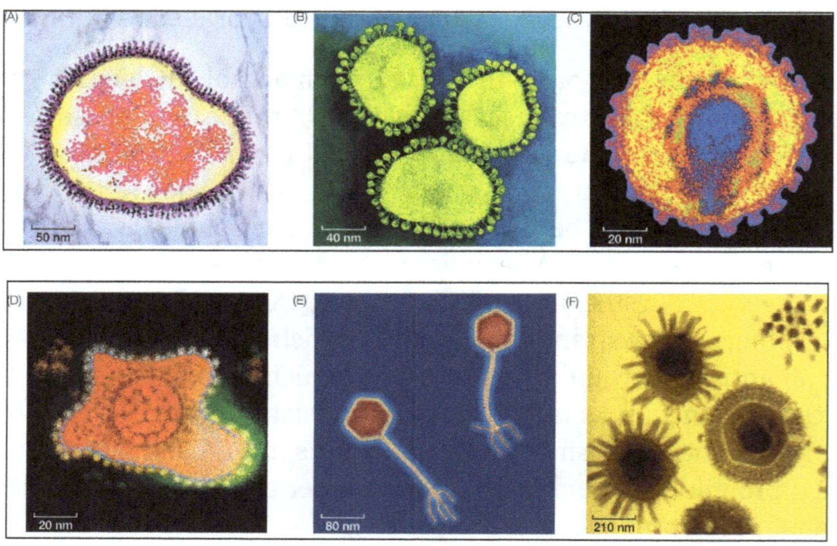

Single-stranded linear DNA, 4–6 kb in size, is found with the members of the Parvovirus family that comprises the parvo-, the erythro- and the dependoviruses. The virion contains 2–4 structural protein species which are differently derived from the same gene product. The adeno-associated virus (AAV, a dependovirus) is incapable of producing progeny virions except in the presence of helper viruses (adenovirus or herpesvirus). It is therefore said to be replication defective.

Circular single-stranded DNA of only 1.7 to 2.3 kb is found in members of the Circovirus family which comprise the smallest autonomously propagated viruses. The isometric capsid measures 17 nm and is composed of 2 protein species only.

Virus Classification on the basis of Shared Properties

On the basis of shared properties viruses are grouped at different hierarchical levels of order, family, subfamily, genus and species. More than 30,000 different virus isolates are known today and grouped in more than 3,600 species, in 164 genera and 71 families. Viral morphology provides the basis for grouping viruses into families. A virus family may consist of members that replicate only in vertebrates, only in invertebrates, only in plants, or only in bacteria. Certain families contain viruses that replicate in more than one of these hosts.

Besides physical properties, several factors pertaining to the mode of replication play a role in classification: the configuration of the nucleic acid (ss or ds, linear or circular), whether the genome consists of one molecule of nucleic acid or is segmented, and whether the strand of ss RNA is sense or antisense. Also considered in classification is the site of viral capsid assembly and, in enveloped viruses, the site of nucleocapsid envelopment.

The use of Latinized names ending in -viridae for virus families and ending in -virus for viral genera has gained wide acceptance. The names of subfamilies end in -virinae. Vernacular names continue to be used to describe the viruses within a genus. In this text, Latinized endings for families and subfamilies usually are not used.

In the early days of virology, viruses were named according to common pathogenic properties, e.g. organ tropism and modes of transmission, and often also after their discoverers. From the early 1950s until the mid-1960s, when many new viruses were being discovered, it was popular to compose virus names by using sigla (abbreviations derived from a few or initial letters). Thus the name Picornaviridae is derived from pico (small) and RNA; the name Reoviridae is derived from respiratory, enteric, and orphan viruses because the agents were found in both respiratory and enteric specimens and were not related to other classified viruses; Papovaviridae is from papilloma, polyoma, and vacuolating agent (simian virus 40 [SV40]); retrovirus is from reverse transcriptase; Hepadnaviridae is from the replication of the virus in hepatocytes and their DNA genomes, as seen in hepatitis B virus. Hepatitis A virus is classified now in the family Picornaviridae, genus Hepatovirus. Although the current rules for nomenclature do not prohibit the introduction of new sigla, they require that the siglum be meaningful to workers in the field and be recognized by international study groups.

The names of the other families that contain viruses pathogenic for humans are derived as follows: Adenoviridae (adeno, "gland"; refers to the adenoid tissue from which the viruses were first isolated); Astroviridae (astron means star); Arenaviridae (arena "sand") describes the sandy appearance of the virion. Bunyaviridae (from Bunyamwera, the place in Africa where the type strain was isolated); Calicivirus (calix, "cup" or "goblet" from the cup-shaped depressions on the viral surfaces); Coronaviridae (corona, "crown") describes the appearance of the peplomers protruding from the viral surface; Filoviridae describes the morphology of these viruses. Herpesviridae (herpes, "creeping") describes the nature of the lesions; Orthomyxoviridae (ortho, "true," plus myxo "mucus," a substance for which the viruses have an affinity; Paramyxoviridae derived from para, "closely resembling" and myxo; Parvoviridae (parvus means, "small"); Poxviridae (pock means, "pustule"); Rhabdoviridae (rhabdo, "rod" describes the shape of the viruses and Togaviridae (toga, "cloak") refers to the tight viral envelope.

Several viruses of medical importance still remain unclassified. Some are difficult or impossible to propagate in standard laboratory host systems and thus cannot be obtained in sufficient quantity to permit more precise characterization. Hepatitis E virus, the Norwalk virus and similar agents that cause nonbacterial gastroenteritis in humans are now assigned to the calicivirus family.

The fatal transmissible dementias in humans and other animals (scrapie in sheep and goat; bovine spongiform encephalopathy in cattle, transmissible mink encephalopathy; Kuru, Creutzfeldt-Jakob disease, and Gerstmann-Straussler-Scheinker syndrome in humans) are caused by the accumulation of non-soluble amyloid fibrils in the central nervous systems. The agents causing transmissible subacute spongiform encephalopathies have been linked to viroids or virinos (i.e. plant pathogens consisting of naked, but very stable circular RNA molecules of about 3-400 bases in size, or infectious genomes enwrapped into a host cell coat) because of their resistance to chemical and physical agents. According to an alternative theory, the term "prion" has been coined to point to an essential nonviral infectious cause for these fatal encephalopathies—prion standing for self-replicating proteinaceous agent devoid of demonstrable nucleic acid. Some of the transmissible amyloidoses show a familial pattern and can be explained by defined mutations which render a primary soluble glycoprotein insoluble, which in turn leads to the pathognomonic accumulation of amyloid fibers and plaques.

Baltimore Classification

The Baltimore classification, developed by David Baltimore, is a virus classification system that groups viruses into families, depending on their type of genome (DNA, RNA, single-stranded (ss), double-stranded (ds), etc.) and their method of replication.

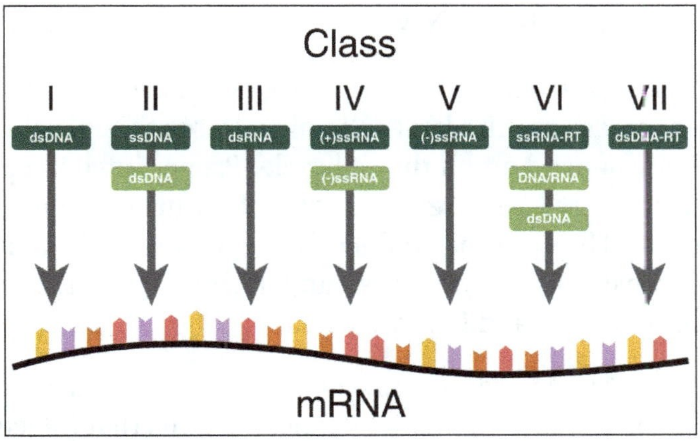

The Baltimore Classification of viruses, created by the American biologist David Baltimore, is based on the method of viral mRNA synthesis.

Classifications

Classifying viruses according to their genome means that those in a given category will all behave in much the same way, which offers some indication of how to proceed with further research. Seven Baltimore classes:

- I: dsDNA viruses (e.g. Adenoviruses, Herpesviruses, Poxviruses).
- II: ssDNA viruses (+ strand or "sense") DNA (e.g. Parvoviruses).
- III: dsRNA viruses (e.g. Reoviruses).
- IV: (+)ssRNA viruses (+ strand or sense) RNA (e.g. Picornaviruses, Togaviruses).
- V: (−)ssRNA viruses (− strand or antisense) RNA (e.g. Orthomyxoviruses, Rhabdoviruses).
- VI: ssRNA-RT viruses (+ strand or sense) RNA with DNA intermediate in life-cycle (e.g. Retroviruses).
- VII: dsDNA-RT viruses DNA with RNA intermediate in life-cycle (e.g. Hepadnaviruses).

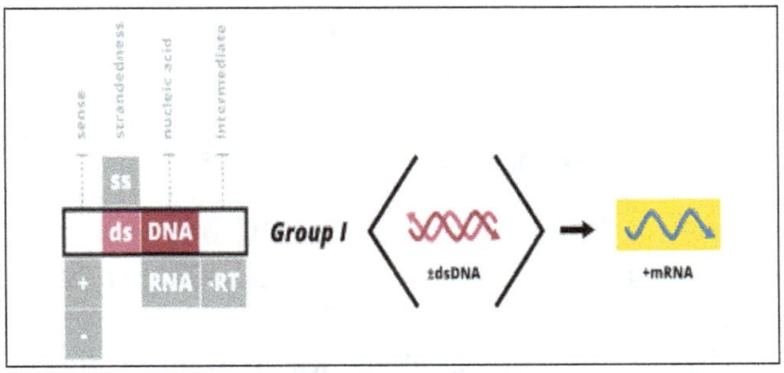

Visualization of the 7 groups of virus according to the Baltimore Classification.

Group I: Double-stranded DNA Viruses

These types of viruses must enter the host nucleus before they are able to replicate. Furthermore, these viruses require host cell polymerases to replicate the viral genome and, hence, are highly dependent on the cell cycle. Proper infection and production of progeny requires that the cell be in replication, as it is during replication that the cell's polymerases are active. The virus may induce the cell to forcefully undergo cell division, which may lead to transformation of the cell and, ultimately, cancer. Examples include Herpesviridae, Adenoviridae, and Papovaviridae.

There is only one well-studied example in which a class 1 virus is not replicating within the nucleus: the Poxvirus family, a highly pathogenic virus that infects vertebrates and includes the smallpox virus.

The mRNA is transcribed regularly from viral DNA using the host's RNA polymerase II. This produces two types of mRNA's: 1) early mRNA, transcribed prior to the synthesis of viral DNA, and 2) late mRNA, transcribed from progeny DNA.

Group II: Single-stranded DNA Viruses

Viruses in this category include the Anelloviridae, Circoviridae, and Parvoviridae (which infect vertebrates), the Geminiviridae and Nanoviridae (which infect plants), and the Microviridae (which infect prokaryotes). Most of them have circular genomes (the parvoviruses are the only known exception). Eukaryote-infecting viruses replicate mostly within the nucleus - usually via a rolling circle mechanism, forming double-stranded DNA intermediate in the process. A prevalent but asymptomatic human Anellovirus, called Transfusion Transmitted Virus (TTV), is included within this classification.

Group III: Double-stranded RNA Viruses

As with most RNA viruses, this class replicates in the "Core" capsid that is in cytoplasm, not having to use the host replication polymerases to as much a degree as DNA viruses. This family is also not as well-studied as the rest and includes 2 major families, the Reoviridae and Birnaviridae. Replication is monocistronic and includes individual, segmented genomes, meaning that each of the genes codes for only one protein, unlike other viruses that exhibit more complex translation.

Group IV and V: Single-stranded RNA Viruses

The ssRNA viruses belong to Class IV or V of the Baltimore classification. They could be grouped into positive sense or negative sense according to the sense of polarity of RNA. The single stranded RNA is the common feature of these viruses. The replication of viruses happens in the cytoplasm or nucleus (for segmented class V viruses). Class IV and V ssRNA viruses do not depend as heavily as DNA viruses on the cell cycle.

Group IV: Single-stranded RNA Viruses - Positive-sense

The positive-sense RNA viruses and indeed all RNA defined as positive-sense can be directly accessed by the host ribosomes immediately to form proteins. These can be divided into two groups, both of which reproduce in the cytoplasm:

- Viruses with polycistronic mRNA where the genome RNA forms the mRNA and is translated into a polyprotein product that is subsequently cleaved to form the mature proteins. This means that the gene can utilize a few methods in which to produce proteins from the same strand of RNA, all in the sake of reducing the size of its gene.
- Viruses with complex transcription, for which subgenomic mRNAs, ribosomal frameshifting, and proteolytic processing of polyproteins may be used. All of which are different mechanisms with which to produce proteins from the same strand of RNA.

Examples of this class include the families Astroviridae, Caliciviridae, Coronaviridae, Flaviviridae, Picornaviridae, Arteriviridae, and Togaviridae.

Group V: Single-stranded RNA Viruses - Negative-sense

The negative-sense RNA viruses and indeed all genes defined as negative-sense cannot be directly accessed by host ribosomes to immediately form proteins. Instead, they must be transcribed by viral polymerases into a "readable" form, which is the positive-sense reciprocal. These can also be divided into two groups:

- Viruses containing nonsegmented genomes for which the first step in replication is transcription from the (-)-stranded genome by the viral RNA-dependent RNA polymerase to yield monocistronic mRNAs that code for the various viral proteins. A positive-sense genome copy is then produced that serves as template for production of the (-)-strand genome. Replication is within the cytoplasm.
- Viruses with segmented genomes for which replication occurs in the nucleus and for which the viral RNA-dependent RNA polymerase produces monocistronic mRNAs from each genome segment. The largest difference between the two is the location of replication.

Examples in this class include the families Arenaviridae, Orthomyxoviridae, Paramyxoviridae, Bunyaviridae, Filoviridae, and Rhabdoviridae (the latter of which includes the rabies virus).

Group VI: Positive-sense Single-stranded RNA Viruses that Replicate through a DNA Intermediate

A well-studied family of this class of viruses include the retroviruses. One defining feature is the use of reverse transcriptase to convert the positive-sense RNA into DNA. Instead of using the RNA for templates of proteins, they use DNA to create the templates, which is spliced into the host genome using integrase. Replication can then commence with the help of the host cell's polymerases.

Group VII: Double-stranded DNA Viruses that Replicate through a Single-stranded RNA Intermediate

This small group of viruses, exemplified by the Hepatitis B virus (which is in the Hepadnaviridae family), have a double-stranded, gapped genome that is subsequently filled in to form a covalently closed circle (cccDNA) that serves as a template for production of viral mRNAs and a subgenomic RNA. The pregenome RNA serves as template for the viral reverse transcriptase for production of the DNA genome.

Animal Virus

Animal viruses are viruses that infect animals. Viruses infect all cellular life and although viruses infect every animal, plant and protist species, each has its own specific range of viruses that often infect only that species.

Vertebrates

The viruses of vertebrates are informally distinguished between those that primarily cause infections of humans and those that infect other animals. The two fields of study are called medical (or clinical) virology and veterinary virology respectively. Although not the first viruses to be discovered and characterised, those that cause infections of humans are the most studied. Different viruses can infect all the organs and tissues of the body and the outcomes range from mild or no symptoms, to life-threatening diseases. Humans cannot be infected by plant or insect viruses, but they are susceptible to infections with viruses from other vertebrates. These are called viral zoonoses or zoonotic infections. Examples include, rabies, yellow fever and pappataci fever.

The viruses that infect other vertebrates are related to those of humans and most families of viruses that cause human diseases are represented. They are important pathogens of livestock and cause diseases such as foot-and-mouth disease and bluetongue. Jersey and Guernsey breeds of cattle are particularly susceptible to pox viruses, with symptoms characterised by widespread, unsightly skin lesions. And most people have heard of myxomatosis, which is a fatal pox virus infection of rabbits: once infected they die within twelve days. The virus was deliberately released in Australia in 1950, in an attempt to control the exponentially growing rabbit population. Rabbits had been brought to the continent in 1859 for sport, and having no natural predators, bred at an extraordinary rate. The infection killed 99.8 percent of rabbits, but by the late 1950s, Australian rabbits started to become immune to the virus and the population of rabbits increased, but never to the vast numbers seen before 1950.

Rabbits around a waterhole during the myxomatosis trial.

Companion animals such as cats, dogs, and horses, if not vaccinated, can catch serious viral infections. Canine parvovirus 2 is caused by a small DNA virus, and infections are often fatal in pups. The emergence of the parvovirus in the 1970s was the most significant in the history of infectious diseases. The disease spread rapidly across the world, and thousands of dogs died from the infection. The virus originated in cats, the vector of feline panleukopenia, but a mutation that changed just two amino acids in the viral capsid protein VP2 allowed it to cross the species barrier, and dogs, unlike cats, had no resistance to the disease. Canine distemper virus is closely related to measles virus and is the most important viral disease of dogs. The disease which was first described in 1760, by Edward Jenner, the pioneer of smallpox vaccination, is highly contagious, but is well controlled by vaccination. In the 1990s, thousands of African lions died from the infection, which they contracted from feral dogs and hyenas.

Marine mammals are susceptible to viral infections. In 1988 and 2002, thousands of harbor seals were killed in Europe by the measles-like phocine distemper virus. Large outbreaks of the disease were recorded among the seal populations of Lake Baikal and along the shores of the Baltic and North Sea. The infection resembled canine distemper; the animals died within two weeks of respiratory distress and many aborted pups were seen. Many other viruses, including caliciviruses, herpesviruses, adenoviruses and parvoviruses, circulate in marine mammal populations.

Fish too have their viruses. They are particularly prone to infections with rhabdoviruses, which are distinct from, but related to rabies virus. At least nine types of rhabdovirus cause economically important diseases in species including salmon, pike, perch, sea bass, carp and cod. The symptoms include anaemia, bleeding, lethargy and a mortality rate that is affected by the temperature of the water. In hatcheries the diseases are often controlled by increasing the temperature to 15–18 °C. Like all vertebrates, fish suffer from herpes viruses. These ancient viruses have co-evolved with their hosts and are highly species-specific. In fish, they cause cancerous tumours and non-cancerous growths called hyperplasia.

Invertebrates

Arthropods is the largest group of animals and has shown to be a major reservoir of different viruses, both insect-specific viruses (ISV) and viruses that can infect both vertebrates and invertebrates, more known as arthropod-borne viruses (arboviruses). Insect-specific viruses are, as the name reveals, characterised by their incapacity to infect vertebrates. This can be assessed through, viral inoculation of mammalian, avian, or amphibian cell lines. The first (ISV) was discovered over 40 years ago by Stollar and Thomas. It was isolated from an Aedes aegypti cell culture where a large number of syncytia were observed and the virus was named cell fusing agent virus (CFAV). Further, when inoculated on different vertebrate cell lines no cytopathic effect (CPE) could be observed and the virus could not be re-isolated, suggesting that the virus must be insect-specific.

Honey bee infected with deformed wing virus.

Invertebrates do not produce antibodies by the lymphocyte-based adaptive immune system that is central to vertebrate immunity, but they are capable of effective immune responses. Phagocytosis was first observed in invertebrates, and this and other innate immune responses are important in immunity to viruses and other pathogens. The hemolymph of invertebrates contains many soluble defence molecules, such as hemocyanins, lectins, and proteins, which protect these animals against invaders.

The health of the honey bee has been important to human societies for centuries. Like all invertebrates, the honey bee (Apis mellifera) is susceptible to many viral infections, and their numbers have dramatically declined around the world. These bees often suffer infestations of varroa mites, which are vectors for deformed wing virus, as a result, this virus has become one of the most widely distributed and contagious insect viruses on the planet. The virus causes stunted wings and as a result, the infected bees are unable to leave the hive and forage for nectar. Symptomatic bees have a severely reduced life-span of less than 48 hours and are often expelled from the hive by other bees. Bees are crucial to the survival of humans, along with producing honey, they pollinate plants that contribute up to one third of the food we eat, and their dramatic decline is a grave concern.

Baculoviruses are among the best studied of the invertebrate viruses. They infect and kill several species of agricultural pests, and as natural insecticides, they have been used to control insect populations in Brazil and Paraguay such as the velvet bean caterpillar (Anticarsia gemmatalis), a pest of soy beans. Viruses are an attractive alternative to chemical pesticides because they are safe to other wildlife and leave no residues.

Viruses can also change the behaviour of their insect hosts to their own advantage. A baculovirus of the gypsy moth (Lymantria dispar) makes their caterpillars climb to the tops of trees where they die. In doing so, they release a shower of millions of progeny viruses that go on to infect more caterpillars.

Foot-and-mouth Disease Virus

Foot-and-mouth disease virus (FMDV) is the pathogen that causes foot-and-mouth disease. It is a picornavirus, the prototypical member of the genus Aphthovirus. The disease, which causes vesicles (blisters) in the mouth and feet of bovids, suids, ovids, caprids and other cloven-hoofed animals is highly infectious and a major plague of animal farming.

The virus particle (25-30 nm) has an icosahedral capsid made of protein, without envelope, containing a single strand of ribonucleic acid (RNA) containing a positive encoding of its genome. When the virus comes in contact with the membrane of a host cell, it binds to a receptor site and triggers a folding-in of the membrane. Once the virus is inside the host cell, the capsid dissolves, and the RNA gets replicated, and translated into viral proteins by the cell's ribosomes using a cap-independent mechanism driven by the internal ribosome entry site element.

The synthesis of viral proteins include 2A 'cleavage' during translation. They include proteases that inhibit the synthesis of normal cell proteins, and other proteins that interact with different components of the host cell. The infected cell ends up producing large quantities of viral RNA and capsid proteins, which are assembled to form new viruses. After assembly, the host cell lyses (bursts) and releases the new viruses.

Foot-and-mouth disease virus occurs in seven major serotypes: O, A, C, SAT-1, SAT-2, SAT-3, and Asia-1. These serotypes show some regionality, and the O serotype is most common.

African Swine Fever Virus

African swine fever virus (ASFV) is a large, double-stranded DNA virus in the Asfarviridae family. It is the causative agent of African swine fever (ASF). The virus causes a hemorrhagic fever with high mortality rates in domestic pigs; some isolates can cause death of animals as quickly as a week after infection. It persistently infects its natural hosts, warthogs, bushpigs, and soft ticks of the genus Ornithodoros, which likely act as a vector, with no disease signs. It does not cause disease in humans. ASFV is endemic to sub-Saharan Africa and exists in the wild through a cycle of infection between ticks and wild pigs, bushpigs, and warthogs. The disease was first described after European settlers brought pigs into areas endemic with ASFV, and as such, is an example of an emerging infectious disease.

ASFV replicates in the cytoplasm of infected cells. It is the only virus with a double-stranded DNA genome known to be transmitted by arthropods.

Virology

ASFV is a large, icosahedral, double-stranded DNA virus with a linear genome of 189 kilobases containing more than 180 genes. The number of genes differs slightly among different isolates of the virus. ASFV has similarities to the other large DNA viruses, e.g., poxvirus, iridovirus, and mimivirus. In common with other viral hemorrhagic fevers, the main target cells for replication are those of monocyte, macrophage lineage. Entry of the virus into the host cell is receptor-mediated, but the precise mechanism of endocytosis is presently unclear.

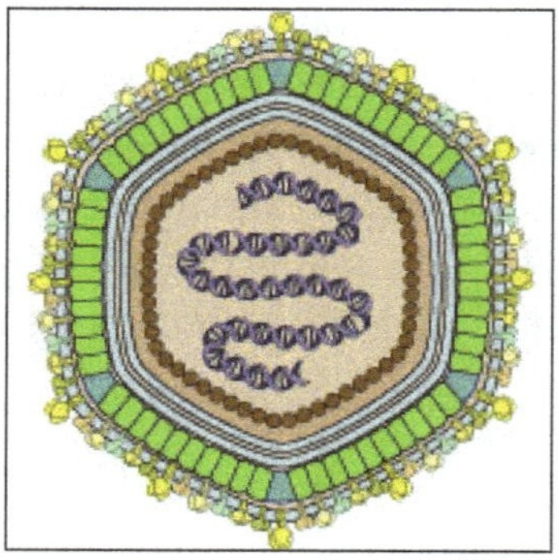

Diagram of ASFV.

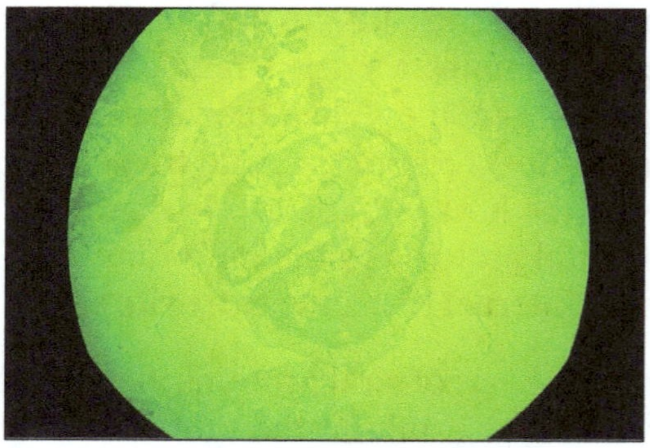

Macrophage cell in early stages of infection with ASFV.

The virus encodes enzymes required for replication and transcription of its genome, including elements of a base excision repair system, structural proteins, and many proteins that are not essential for replication in cells, but instead have roles in virus survival and transmission in its hosts. Virus replication takes place in perinuclear factory areas. It is a highly orchestrated process with at least four stages of transcription—immediate-early, early, intermediate, and late. The majority of replication and assembly occurs in discrete, perinuclear regions of the cell called virus factories, and finally progeny virions are transported to the plasma membrane along microtubules where they bud out or are propelled away along actin projections to infect new cells. As the virus progresses through its lifecycle, most if not all of the host cell's organelles are modified, adapted, or in some cases destroyed.

Assembly of the icosahedral capsid occurs on modified membranes from the endoplasmic reticulum. Products from proteolytically processed polyproteins form the core shell between the internal membrane and the nucleoprotein core. An additional outer membrane is gained as particles bud from the plasma membrane. The virus encodes proteins that inhibit signalling pathways in infected macrophages and thus modulate transcriptional activation of immune response genes. In addition, the virus encodes proteins which inhibit apoptosis of infected cells to facilitate production of progeny virions. Viral membrane proteins with similarity to cellular adhesion proteins modulate interaction of virus-infected cells and extracellular virions with host components.

Genotypes

Based on sequence variation in the C-terminal region of the B646L gene encoding the major capsid protein p72, 22 ASFV genotypes (I–XXIII) have been identified. All ASFV p72 genotypes have been circulating in eastern and southern Africa. Genotype I has been circulating in Europe, South America, the Caribbean, and western Africa. Genotype VIII is confined to four East African countries.

Signs and Symptoms

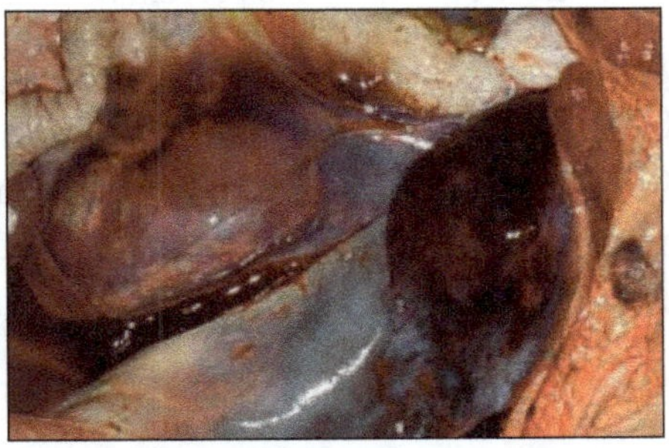

The swelling around the kidneys and the muscle hemorrhages visible here are typical of pigs with African swine fever.

In the acute form of the disease caused by highly virulent strains, pigs may develop a high fever, but show no other noticeable symptoms for the first few days. They then gradually lose their appetites and become depressed. In white-skinned pigs, the extremities turn blueish-purple and hemorrhages become apparent on the ears and abdomen. Groups of infected pigs lie huddled together shivering, breathing abnormally, and sometimes coughing. If forced to stand, they appear unsteady on their legs; this is called congenital tremor type A-I in newborn piglets. Within a few days of infection, they enter a comatose state and then die. In pregnant sows, spontaneous abortions occur.

In milder infections, affected pigs lose weight, become thin, and develop signs of pneumonia, skin ulcers, and swollen joints.

Diagnosis

The clinical symptoms of ASFV infection are very similar to classical swine fever, and the two diseases normally have to be distinguished by laboratory diagnosis. This diagnosis is usually performed by an ELISA or isolation of the virus from either the blood, lymph nodes, spleen, or serum of an infected pig.

Spread

The virus can be spread by ticks, but also by swine eating pork products that contain the virus.

A multi-language sign in Germany telling people to dispose of waste food in closed waste bins to prevent boars eating it.

The National Pig Association, a UK industry body, states that the virus can also be transmitted by direct or indirect contact with infected pigs, faeces or body fluids. As the virus may survive 11 days in pig faeces, and months or years in pork products, the association advises strict biosecurity measures for pig farms including a three-day quarantine on entering the UK, and avoiding both pigs and areas where wild boar are found.

Vaccine Research

As of 2020, no vaccine is commercially available. Immunity appears to be largely strain specific. An experimental vaccine is in development against the 2007 Georgia isolate currently circulating, which is attenuated by deletion of the viral I177L gene.

Canine Parvovirus

Canine parvovirus (also referred to as CPV, CPV2, or parvo) is a contagious virus mainly affecting dogs. CPV is highly contagious and is spread from dog to dog by direct or indirect contact with their feces. Vaccines can prevent this infection, but mortality can reach 91% in untreated cases. Treatment often involves veterinary hospitalization. Canine parvovirus may infect other mammals including foxes, wolves, cats, and skunks.

Signs

Dogs that develop the disease show signs of the illness within 3 to 7 days. The signs may include lethargy, vomiting, fever, and diarrhea (usually bloody). Generally, the first sign of CPV is lethargy. Secondary signs are loss of weight and appetite or diarrhea followed by vomiting. Diarrhea and vomiting result in dehydration that upsets the electrolyte balance and this may affect the dog critically. Secondary infections occur as a result of the weakened immune system. Because the normal intestinal lining is also compromised, blood and protein leak into the intestines, leading to anemia and loss of protein, and endotoxins escape into the bloodstream, causing endotoxemia. Dogs have a distinctive odor in the later stages of the infection. The white blood cell level falls, further weakening the dog. Any or all of these factors can lead to shock and death.

Diagnosis

Diagnosis is made through detection of CPV2 in the feces by either an ELISA or a hemagglutination test, or by electron microscopy. PCR has become available to diagnose CPV2, and can be used later in the disease when potentially less virus is being shed in the feces that may not be detectable by ELISA. Clinically, the intestinal form of the infection can sometimes be confused with coronavirus or other forms of enteritis. Parvovirus, however, is more serious and the presence of bloody diarrhea, a low white blood cell count, and necrosis of the intestinal lining also point more towards parvovirus, especially in an unvaccinated dog. The cardiac form is typically easier to diagnose because the symptoms are distinct.

Treatment

Survival rate depends on how quickly CPV is diagnosed, the age of the dog, and how aggressive the treatment is. Treatment usually involves extensive hospitalization, due to severe dehydration and potential damage to the intestines and bone marrow. A CPV test should be given as early as possible if CPV is suspected in order to begin early treatment and increase survival rate if the disease is found.

Treatment ideally also consists of crystalloid IV fluids and colloids (e.g., Hetastarch), antinausea injections (antiemetics) such as maropitant, metoclopramide, dolasetron, ondansetron and prochlorperazine, and broad-spectrum antibiotic injections such as cefazolin/enrofloxacin, ampicillin/enrofloxacin, metronidazole, timentin, or enrofloxacin. IV fluids are administered and antinausea and antibiotic injections are given subcutaneously, intramuscularly, or intravenously. The fluids are typically a mix of a sterile, balanced electrolyte solution, with an appropriate amount of B-complex vitamins, dextrose, and potassium chloride. Analgesic medications can be used to counteract the intestinal discomfort caused by frequent bouts of diarrhea; however, the use of opioid analgesics can result in secondary ileus and decreased motility.

In addition to fluids given to achieve adequate rehydration, each time the puppy vomits or has diarrhea in a significant quantity, an equal amount of fluid is administered intravenously. The fluid requirements of a patient are determined by the animal's body weight, weight changes over time, degree of dehydration at presentation, and surface area.

A blood plasma transfusion from a donor dog that has already survived CPV is sometimes used to provide passive immunity to the sick dog. Some veterinarians keep these dogs on site, or have frozen serum available. There have been no controlled studies regarding this treatment. Additionally, fresh frozen plasma and human albumin transfusions can help replace the extreme protein losses seen in severe cases and help assure adequate tissue healing. However, this is controversial with the availability of safer colloids such as Hetastarch, as it will also increase the colloid osmotic pressure without the ill effect of predisposing that canine patient to future transfusion reaction.

Once the dog can keep fluids down, the IV fluids are gradually discontinued, and very bland food slowly introduced. Oral antibiotics are administered for a number of days depending on the white blood cell count and the patient's ability to fight off secondary infection. A puppy with minimal symptoms can recover in 2 or 3 days if the IV fluids are begun as soon as symptoms are noticed and the CPV test confirms the diagnosis. If more severe, depending on treatment, puppies can remain ill from 5 days up to 2 weeks. However, even with hospitalization, there is no guarantee that the dog will be cured and survive.

Unconventional Treatments

There have been anecdotal reports of oseltamivir (Tamiflu) reducing disease severity and hospitalization time in canine parvovirus infection. The drug may limit the ability of the virus to invade the crypt cells of the small intestine and decrease gastrointestinal bacteria colonization and toxin production. However, due to the viral DNA replication pattern of parvovirus and the mechanism of action of oseltamivir, this medication has not shown to improve survival rates or shorten hospitalization stay. Lastly, recombinant feline interferon omega (rFeIFN-ω), produced in silkworm larvae using a baculovirus vector, has been demonstrated by multiple studies to be an effective treatment. However, this therapy is not currently approved in the United States.

An unpublished 2012 study from Colorado State University showed good results with an intensive at-home treatment using maropitant (Cerenia) and Convenia (a long acting antibiotic injection), two drugs released by Zoetis (formerly Pfizer). This treatment was based on outpatient care, and would cost $200 to $300, a fraction of the $1,500 to $3,000 that inpatient care cost. However, the more-effective care is intravenous (IV) fluid therapy. In the CSU study, survival rate for the new treatment group was 85%, compared to the 90% survival for the conventional inpatient treatment. Note that the outpatient dogs received initial intravenous fluid resuscitation, and had aggressive subcutaneous fluid therapy and daily monitoring by a veterinarian.

Virology

CPV2 is a non-enveloped single-stranded DNA virus in the Parvoviridae family. As the virus is only 20 to 26 nm in diameter. It has an icosahedral symmetry. The genome is about 5000 nucleotides long. CPV2 continues to evolve, and the success of new strains seems to depend on extending the range of hosts affected and improved binding to its receptor, the canine transferrin receptor. CPV2 has a high rate of evolution, possibly due to a rate of nucleotide substitution that is more like RNA viruses such as Influenzavirus A. In contrast, FPV seems to evolve only through random genetic drift.

CPV2 affects dogs, wolves, foxes, and other canids. CPV2a and CPV2b have been isolated from a small percentage of symptomatic cats and is more common than feline panleukopenia in big cats.

Previously it has been thought that the virus does not undergo cross species infection. However studies in Vietnam have shown that CPV2 can undergo minor antigenic shift and natural mutation to infect felids. Analyses of feline parvovirus (FPV) isolates in Vietnam and Taiwan revealed that more than 80% of the isolates were of the canine parvovirus type, rather than feline panleukopenia virus (FPLV). CPV2 may spread to cats easier than dogs and undergo faster rates of mutation within that species.

Prevention and Decontamination

Prevention is the only way to ensure that a puppy or dog remains healthy because the disease is extremely virulent and contagious. Appropriate vaccination should be performed starting at 7–8 weeks of age, with a booster given every 3–4 weeks until at least 16 weeks of age. Pregnant mothers should not be vaccinated as it will abort the puppies and could make the mother extremely sick. The virus is extremely hardy and has been found to survive in feces and other organic material such as soil for up to 1 year. It survives extremely low and high temperatures. The only household disinfectant that kills the virus is bleach. The dilute bleach solution needs to be in a 1:10 ratio to disinfect and kill parvovirus.

Puppies are generally vaccinated in a series of doses, extending from the earliest time that the immunity derived from the mother wears off until after that passive immunity is definitely gone. Older puppies (16 weeks or older) are given 3 vaccinations 3 to 4 weeks apart. The duration of immunity of vaccines for CPV2 has been tested for all major vaccine manufacturers in the United States and has been found to be at least three years after the initial puppy series and a booster 1 year later.

A dog that successfully recovers from CPV2 generally remains contagious for up to three weeks, but it is possible they may remain contagious for up to six. Ongoing infection risk is primarily from fecal contamination of the environment due to the virus's ability to survive many months in the environment. Neighbours and family members with dogs should be notified of infected animals so that they can ensure that their dogs are vaccinated or tested for immunity. A modified live vaccine may confer protection in 3 to 5 days; the contagious individual should remain in quarantine until other animals are protected.

Plant Virus

Plant viruses are viruses that affect plants. Like all other viruses, plant viruses are obligate intracellular parasites that do not have the molecular machinery to replicate without a host. Plant viruses can be pathogenic to higher plants.

Pepper mild mottle virus.

Leaf curl virus.

Most plant viruses are rod-shaped, with protein discs forming a tube surrounding the viral genome; isometric particles are another common structure. They rarely have an envelope. The great majority have an RNA genome, which is usually small and single stranded (ss), but some viruses have double-stranded (ds) RNA, ssDNA or dsDNA genomes. Although plant viruses are not as well understood as their animal counterparts, one plant virus has become iconic: tobacco mosaic virus (TMV), the first virus to be discovered. This and other viruses cause an estimated U.S$60 billion loss in crop yields worldwide each year. Plant viruses are grouped into 73 genera and 49 families. However, these figures relate only to cultivated plants, which represent only a tiny fraction of the total number of plant species. Viruses in wild plants have been relatively little studied, but the interactions between wild plants and their viruses often do not appear to cause disease in the host plants.

To transmit from one plant to another and from one plant cell to another, plant viruses must use strategies that are usually different from animal viruses. Plants do not move, and so plant-to-plant transmission usually involves vectors (such as insects). Plant cells are surrounded by solid cell walls, therefore transport through plasmodesmata is the preferred path for virions to move between plant cells. Plants have specialized mechanisms for transporting mRNAs through plasmodesmata, and these mechanisms are thought to be used by RNA viruses to spread from one cell to another. Plant defenses against viral infection include, among other measures, the use of siRNA in response to dsRNA. Most plant viruses encode a protein to suppress this response. Plants also reduce transport through plasmodesmata in response to injury.

The discovery of plant viruses causing disease is often accredited to A. Mayer working in the Netherlands demonstrated that the sap of mosaic obtained from tobacco leaves developed mosaic symptom when injected in healthy plants. However the infection of the sap was destroyed when it was boiled. He thought that the causal agent was the bacteria. However, after larger inoculation with a large number of bacteria, he failed to develop a mosaic symptom.

In 1898, Martinus Beijerinck, who was a Professor of Microbiology at the Technical University the Netherlands, put forth his concepts that viruses were small and determined that the "mosaic disease" remained infectious when passed through a Chamberland filter-candle. This was in contrast to bacteria microorganisms, which were retained by the filter. Beijerinck referred to the infectious filtrate as a "contagium vivum fluidum", thus the coinage of the modern term "virus".

After the initial discovery of the 'viral concept' there was need to classify any other known viral diseases based on the mode of transmission even though microscopic observation proved fruitless. In 1939 Holmes published a classification list of 129 plant viruses. This was expanded and in 1999 there were 977 officially recognized, and some provisional, plant virus species.

The purification (crystallization) of TMV was first performed by Wendell Stanley, who published his findings in 1935, although he did not determine that the RNA was the infectious material. However, he received the Nobel Prize in Chemistry in 1946. In the 1950s a discovery by two labs simultaneously proved that the purified RNA of the TMV was infectious which reinforced the argument. The RNA carries genetic information to code for the production of new infectious particles.

More recently virus research has been focused on understanding the genetics and molecular biology of plant virus genomes, with a particular interest in determining how the virus can replicate, move and infect plants. Understanding the virus genetics and protein functions has been used to explore the potential for commercial use by biotechnology companies. In particular, viral-derived sequences have been used to provide an understanding of novel forms of resistance. The recent boom in technology allowing humans to manipulate plant viruses may provide new strategies for production of value-added proteins in plants.

Structure

Viruses are extremely small and can only be observed under an electron microscope. The structure of a virus is given by its coat of proteins, which surround the viral genome. Assembly of viral particles takes place spontaneously.

Over 50% of known plant viruses are rod-shaped (flexuous or rigid). The length of the particle is normally dependent on the genome but it is usually between 300–500 nm with a diameter of 15–20 nm. Protein subunits can be placed around the circumference of a circle to form a disc. In the presence of the viral genome, the discs are stacked, then a tube is created with room for the nucleic acid genome in the middle.

The second most common structure amongst plant viruses are isometric particles. They are 25–50 nm in diameter. In cases when there is only a single coat protein, the basic structure consists of 60 T subunits, where T is an integer. Some viruses may have 2 coat proteins that associate to form an icosahedral shaped particle.

There are three genera of Geminiviridae that consist of particles that are like two isometric particles stuck together.

A very small number of plant viruses have, in addition to their coat proteins, a lipid envelope. This is derived from the plant cell membrane as the virus particle buds off from the cell.

Transmission of Plant Viruses

Through Sap

Viruses can be spread by direct transfer of sap by contact of a wounded plant with a healthy one. Such contact may occur during agricultural practices, as by damage caused by tools or hands, or naturally, as by an animal feeding on the plant. Generally TMV, potato viruses and cucumber mosaic viruses are transmitted via sap.

Insects

Plant viruses need to be transmitted by a vector, most often insects such as leafhoppers. One class of viruses, the Rhabdoviridae, has been proposed to actually be insect viruses that have evolved to replicate in plants. The chosen insect vector of a plant virus will often be the determining factor in that virus's host range: it can only infect plants that the insect vector feeds upon. This was shown in part when the old world white fly made it to the United States, where it transferred many plant viruses into new hosts. Depending on the way they are transmitted, plant viruses are classified as non-persistent, semi-persistent and persistent. In non-persistent transmission, viruses become attached to the distal tip of the stylet of the insect and on the next plant it feeds on, it inoculates it with the virus. Semi-persistent viral transmission involves the virus entering the foregut of the insect. Those viruses that manage to pass through the gut into the haemolymph and then to the salivary glands are known as persistent.

There are two sub-classes of persistent viruses: propagative and circulative. Propagative viruses are able to replicate in both the plant and the insect (and may have originally been insect viruses), whereas circulative can not. Circulative viruses are protected inside aphids by the chaperone protein symbionin, produced by bacterial symbionts. Many plant viruses encode within their genome polypeptides with domains essential for transmission by insects. In non-persistent and semi-persistent viruses, these domains are in the coat protein and another protein known as the helper component. A bridging hypothesis has been proposed to explain how these proteins aid in insect-mediated viral transmission. The helper component will bind to the specific domain of the coat protein, and then the insect mouthparts – creating a bridge. In persistent propagative viruses, such as tomato spotted wilt virus (TSWV), there is often a lipid coat surrounding the proteins that is not seen in other classes of plant viruses. In the case of TSWV, 2 viral proteins are expressed in this lipid envelope. It has been proposed that the viruses bind via these proteins and are then taken into the insect cell by receptor-mediated endocytosis.

Nematodes

Soil-borne nematodes also have been shown to transmit viruses. They acquire and transmit them by feeding on infected roots. Viruses can be transmitted both non-persistently and persistently, but there is no evidence of viruses being able to replicate in nematodes. The virions attach to the stylet (feeding organ) or to the gut when they feed on an infected plant and can then detach during later feeding to infect other plants. Examples of viruses that can be transmitted by nematodes include tobacco ringspot virus and tobacco rattle virus.

Plasmodiophorids

A number of virus genera are transmitted, both persistently and non-persistently, by soil borne zoosporic protozoa. These protozoa are not phytopathogenic themselves, but parasitic. Transmission of the virus takes place when they become associated with the plant roots. Examples include Polymyxa graminis, which has been shown to transmit plant viral diseases in cereal crops and Polymyxa betae which transmits Beet necrotic yellow vein virus. Plasmodiophorids also create wounds in the plant's root through which other viruses can enter.

Seed and Pollen Borne Viruses

Plant virus transmission from generation to generation occurs in about 20% of plant viruses. When viruses are transmitted by seeds, the seed is infected in the generative cells and the virus is maintained in the germ cells and sometimes, but less often, in the seed coat. When the growth and development of plants is delayed because of situations like unfavorable weather, there is an increase in the amount of virus infections in seeds. There does not seem to be a correlation between the location of the seed on the plant and its chances of being infected. Little is known about the mechanisms involved in the transmission of plant viruses via seeds, although it is known that it is environmentally influenced and that seed transmission occurs because of a direct invasion of the embryo via the ovule or by an indirect route with an attack on the embryo mediated by infected gametes. These processes can occur concurrently or separately depending on the host plant. It is unknown how the virus is able to directly invade and cross the embryo and boundary between the parental and progeny generations in the ovule. Many plants species can be infected through seeds including but not limited to the families Leguminosae, Solanaceae, Compositae, Rosaceae, Cucurbitaceae, Gramineae. Bean common mosaic virus is transmitted through seeds.

Direct Plant-to-human Transmission

Researchers have found tenuous evidence that suggest a virus common to peppers, the Pepper Mild Mottle Virus (PMMoV) may have moved on to infect humans. This is a very rare and highly unlikely event as, to enter a cell and replicate, a virus must "bind to a receptor on its surface, and a plant virus would be highly unlikely to recognize a receptor on a human cell. One possibility is that the virus does not infect human cells directly. Instead, the naked viral RNA may alter the function of the cells through a mechanism similar to RNA interference, in which the presence of certain RNA sequences can turn genes on and off," according to Virologist Robert Garry from the Tulane University in New Orleans, Louisiana.

Translation of Plant Viral Proteins

75% of plant viruses have genomes that consist of single stranded RNA (ssRNA). 65% of plant viruses have +ssRNA, meaning that they are in the same sense orientation as messenger RNA but 10% have -ssRNA, meaning they must be converted to +ssRNA before they can be translated. 5% are double stranded RNA and so can be immediately translated as +ssRNA viruses. 3% require a reverse transcriptase enzyme to convert between RNA and DNA. 17% of plant viruses are ssDNA and very few are dsDNA, in contrast a quarter of animal viruses are dsDNA and three-quarters of bacteriophage are dsDNA. Viruses use the plant ribosomes to produce the 4-10 proteins encoded by their genome. However, since many of the proteins are encoded on a single strand (that is, they are polycistronic) this will mean that the ribosome will either only produce one protein, as it will terminate translation at the first stop codon, or that a polyprotein will be produced. Plant viruses have had to evolve special techniques to allow the production of viral proteins by plant cells.

5' Cap

For translation to occur, eukaryotic mRNAs require a 5' Cap structure. This means that viruses must also have one. This normally consists of 7MeGpppN where N is normally adenine or guanine. The viruses encode a protein, normally a replicase, with a methyltransferase activity to allow this.

Some viruses are cap-snatchers. During this process, a ^{7m}G-capped host mRNA is recruited by the viral transcriptase complex and subsequently cleaved by a virally encoded endonuclease. The resulting capped leader RNA is used to prime transcription on the viral genome.

However some plant viruses do not use cap, yet translate efficiently due to cap-independent translation enhancers present in 5' and 3' untranslated regions of viral mRNA.

Readthrough

Some viruses (e.g. tobacco mosaic virus (TMV)) have RNA sequences that contain a "leaky" stop codon. In TMV 95% of the time the host ribosome will terminate the synthesis of the polypeptide at this codon but the rest of the time it continues past it. This means that 5% of the proteins produced are larger than and different from the others normally produced, which is a form of translational regulation. In TMV, this extra sequence of polypeptide is an RNA polymerase that replicates its genome.

Production of Sub-genomic RNAs

Some viruses use the production of subgenomic RNAs to ensure the translation of all proteins within their genomes. In this process the first protein encoded on the genome, and is the first to be translated, is a replicase. This protein will act on the rest of the genome producing negative strand sub-genomic RNAs then act upon these to form positive strand sub-genomic RNAs that are essentially mRNAs ready for translation.

Segmented Genomes

Some viral families, such as the Bromoviridae instead opt to have multipartite genomes, genomes split between multiple viral particles. For infection to occur, the plant must be infected with all particles across the genome. For instance Brome mosaic virus has a genome split between 3 viral particles, and all 3 particles with the different RNAs are required for infection to take place.

Polyprotein Processing

This strategy is adopted by viral genera such as the Potyviridae and Tymoviridae. The ribosome translates a single protein from the viral genome. Within the polyprotein is an enzyme (or enzymes) with proteinase function that is able to cleave the polyprotein into the various single proteins or just cleave away the protease, which can then cleave other polypeptides producing the mature proteins.

Understanding Plant Viruses

Tobacco mosaic virus (TMV) and Cauliflower mosaic virus (CaMV) are frequently used in plant molecular biology. Of special interest is the CaMV 35S promoter, which is a very strong promoter most frequently used in plant transformations.

Tobacco Mosaic Virus

Tobacco mosaic virus (TMV) is a positive-sense single stranded RNA virus in the genus Tobamovirus that infects a wide range of plants, especially tobacco and other members of the family Solanaceae. The infection causes characteristic patterns, such as "mosaic"-like mottling and discoloration on the leaves (hence the name). TMV was the first virus to be discovered. Although it was known from the late 19th century that a non-bacterial infectious disease was damaging tobacco crops, it was not until 1930 that the infectious agent was determined to be a virus. It is the first pathogen identified as a virus.

In 1886, Adolf Mayer first described the tobacco mosaic disease that could be transferred between plants, similar to bacterial infections. In 1892, Dmitri Ivanovsky gave the first concrete evidence for the existence of a non-bacterial infectious agent, showing that infected sap remained infectious even after filtering through finest Chamberland filters. Later, in 1903, Ivanovsky published a paper describing abnormal crystal intracellular inclusions in the host cells of the affected tobacco plants and argued the connection between these inclusions and the infectious agent. However, Ivanovsky remained rather convinced, despite repeated failures to produce evidence, that the causal agent was an unculturable bacterium, too small to be retained on the employed Chamberland filters and to be detected in the light microscope. In 1898, Martinus Beijerinck independently replicated Ivanovsky's filtration experiments and then showed that the infectious agent was able to reproduce and multiply in the host cells of the tobacco plant. Beijerinck coined the term of "virus" to indicate that the causal agent of tobacco mosaic disease was of non-bacterial nature. Tobacco mosaic virus was the first virus to be crystallized. It was achieved by Wendell Meredith Stanley in 1935 who also showed that TMV remains active even after crystallization. For his work, he was awarded 1/4 of the Nobel Prize in Chemistry in 1946, even though it was later shown some of his conclusions (in particular, that the crystals were pure protein, and assembled by autocatalysis) were incorrect. The first electron microscopical images of TMV were made in 1939 by Gustav Kausche, Edgar Pfankuch and Helmut Ruska – the brother of Nobel Prize winner Ernst Ruska. In 1955, Heinz Fraenkel-Conrat and Robley Williams showed that purified TMV RNA and its capsid (coat) protein assemble by themselves to functional viruses, indicating that this is the most stable structure (the one with the lowest free energy). The crystallographer Rosalind Franklin worked for Stanley for about a month at Berkeley, and later designed and built a model of TMV for the 1958 World's Fair at Brussels. In 1958, she speculated that the virus was hollow, not solid, and hypothesized that the RNA of TMV is single-stranded. This conjecture was proven to be correct after her death and is now known to be the + strand. The investigations of tobacco mosaic disease and subsequent discovery of its viral nature were instrumental in the establishment of the general concepts of virology.

Structure

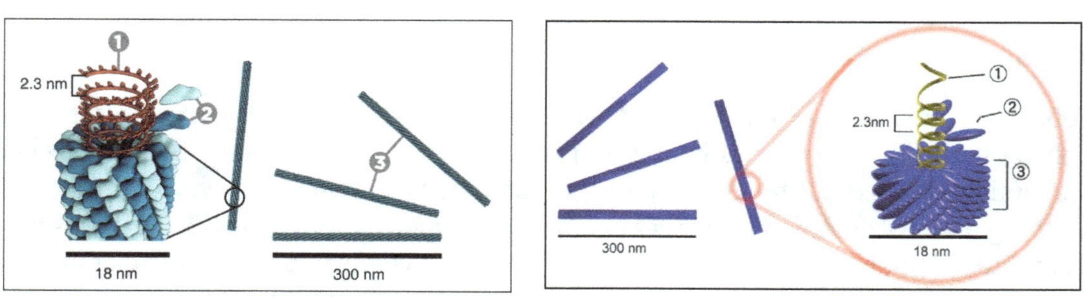

Schematic model of TMV: 1. nucleic acid (RNA), 2. capsomer protein (protomer), 3. capsid.

Tobacco mosaic virus has a rod-like appearance. Its capsid is made from 2130 molecules of coat protein and one molecule of genomic single strand RNA, 6400 bases long. The coat protein self-assembles into the rod-like helical structure (16.3 proteins per helix turn) around the RNA, which forms a hairpin loop structure. The protein monomer consists of 158 amino acids which are assembled into four main alpha-helices, which are joined by a prominent loop proximal to the axis of the virion. Virions are ~300 nm in length and ~18 nm in diameter. Negatively stained electron microphotographs show a distinct inner channel of ~4 nm. The RNA is located at a radius of ~6 nm and is protected from the action of cellular enzymes by the coat protein. There are three RNA nucleotides per protein monomer. X-ray fiber diffraction structure of the intact virus was studied based on an electron density map at 3.6 Å resolution. Inside the core helix, coiled RNA molecule is present, which is made up of nearly 6500 nucleotides.

Genome

The TMV genome consists of a 6.3–6.5 kb single-stranded (ss) RNA. The 3'-terminus has a tRNA-like structure, and the 5' terminus has a methylated nucleotide cap. (m7G5'pppG). The genome encodes 4 open reading frames (ORFs), two of which produce a single protein due to ribosomal readthrough of a leaky UAG stop codon. The 4 genes encode a replicase (with methyltransferase [MT] and RNA helicase [Hel] domains), an RNA-dependent RNA polymerase, a so-called movement protein (MP) and a capsid protein (CP).

Physicochemical Properties

TMV is a thermostable virus. On a dried leaf, it can withstand up to 50 °C (120 degree Fahrenheit) for 30 minutes.

TMV has an index of refraction of about 1.57.

Disease Cycle

TMV does not have a distinct overwintering structure. Rather, it will over-winter in infected tobacco stalks and leaves in the soil, on the surface of contaminated seed (TMV can even survive in contaminated tobacco products for many years). With the direct contact with host plants through its vectors (normally insects such as aphids and leafhoppers), TMV will go through the infection process and then the replication process.

Infection and Transmission

After its multiplication, it enters the neighbouring cells through plasmodesmata. The infection spreads by direct contact to the neighbouring cells. For its smooth entry, TMV produces a 30 kDa movement protein called P30 which enlarges the plasmodesmata. TMV most likely moves from cell-to-cell as a complex of the RNA, P30, and replicate proteins.

It can also spread through phloem for longer distance movement within the plant. Moreover, TMV can be transmitted from one plant to another by direct contact. Although TMV does not have defined transmission vectors, the virus can be easily transmitted from the infected hosts to the healthy plants, by human handling.

Replication

Following entry into its host via mechanical inoculation, TMV uncoats itself to release its viral [+]RNA strand. As uncoating occurs, the MetHel: Pol gene is translated to make the capping enzyme MetHel and the RNA Polymerase. Then the viral genome will further replicate to produce multiple mRNAs via a [-]RNA intermediate primed by the $tRNA_{HIS}$ at the [+]RNA 3' end. The resulting mRNAs encode several proteins, including the coat protein and an RNA-dependent RNA polymerase (RdRp), as well as the movement protein. Thus TMV can replicate its own genome.

After the coat protein and RNA genome of TMV have been synthesized, they spontaneously assemble into complete TMV virions in a highly organized process. The protomers come together to form disks or 'lockwashers' composed of two layers of protomers arranged in a helix. The helical capsid grows by the addition of protomers to the end of the rod. As the rod lengthens, the RNA passes through a channel in its center and forms a loop at the growing end. In this way the RNA can easily fit as a spiral into the interior of the helical capsid.

Host and Symptoms

Tobacco mosaic virus symptoms on tobacco.

Tobacco mosaic virus symptoms on orchid.

Like other plant pathogenic viruses, TMV has a very wide host range and has different effects depending on the host being infected. Tobacco mosaic virus has been known to cause a production loss for flue cured tobacco of up to two percent in North Carolina. It is known to infect members of nine plant families, and at least 125 individual species, including tobacco, tomato, pepper (all members of the useful Solanaceae), cucumbers, and a number of ornamental flowers. There are many different strains. The first symptom of this virus disease is a light green coloration between the veins of young leaves. This is followed quickly by the development of a "mosaic" or mottled pattern of light and dark green areas in the leaves. Rugosity may also be seen where the infected plant leaves display small localized random wrinkles. These symptoms develop quickly and are more pronounced on younger leaves. Its infection does not result in plant death, but if infection occurs early in the season, plants are stunted. Lower leaves are subjected to "mosaic burn" especially during periods of hot and dry weather. In these cases, large dead areas develop in the leaves. This constitutes one of the most destructive phases of

Tobacco mosaic virus infection. Infected leaves may be crinkled, puckered, or elongated. However, if TMV infects crops like grape and apple, it is almost symptomless.

Environment

TMV is known as one of the most stable viruses. It has a very wide survival range. As long as the surrounding temperature remains below approximately 40 degrees Celsius, TMV can sustain its stable form. All it needs is a host to infect. If necessary, greenhouses and botanical gardens would provide the most favorable condition for TMV to spread out, due to the high population density of possible hosts and the constant temperature throughout the year.

Treatment and Management

One of the common control methods for TMV is sanitation, which includes removing infected plants and washing hands in between each planting. Crop rotation should also be employed to avoid infected soil/seed beds for at least two years. As for any plant disease, looking for resistant strains against TMV may also be advised. Furthermore, the cross protection method can be administered, where the stronger strain of TMV infection is inhibited by infecting the host plant with mild strain of TMV, similar to the effect of a vaccine.

In the past ten years, the application of genetic engineering on a host plant genome has been developed to allow the host plant to produce the TMV coat protein within their cells. It was hypothesized that the TMV genome will be re-coated rapidly upon entering the host cell, thus it prevents the initiation of TMV replication. Later it was found that the mechanism that protects the host from viral genome insertion is through gene silencing.

Scientific and Environmental Impact

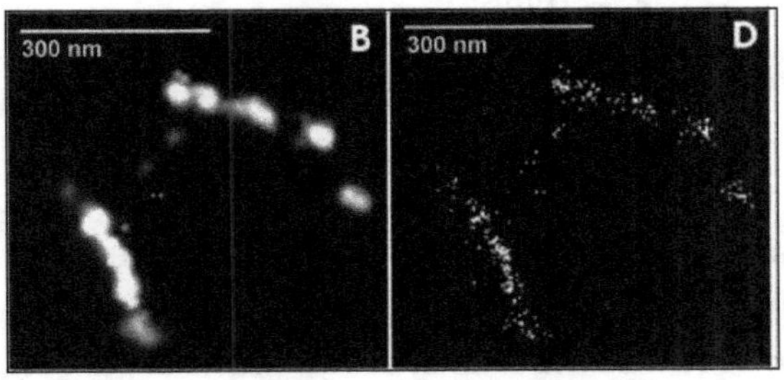

TMV virus: super resolution light microscopy.

The large amount of literature about TMV and its choice for many pioneering investigations in structural biology (including X-ray diffraction), virus assembly and disassembly, and so on, are fundamentally due to the large quantities that can be obtained, plus the fact that it does not infect animals. After growing a few infected tobacco plants in a greenhouse and a few simple laboratory procedures, a scientist can easily produce several grams of the virus.

James D. Watson, in his memoir The Double Helix, cites his x-ray investigation of TMV's helical structure as an important step in deducing the nature of the DNA molecule.

Investigational Uses

Due to its cylindrical shape, high aspect-ratio, self-assembling nature, and ability to incorporate metal coatings (nickel and cobalt) into its shell, TMV is an ideal candidate to be incorporated into battery electrodes. Addition of TMV to a battery electrode increases the reactive surface area by an order of magnitude, resulting in an increase in the battery's capacity by up to six times compared to a planar electrode geometry.

Tomato Spotted Wilt Virus

Tomato spotted wilt virus (TSWV) is a spherical negative-sense RNA virus that has a diameter between 80-110nm.

Transmission and Lifespan

TSWV, which is transmitted by thrips, causes serious losses in economically important crops and it is one of the most economically devastating plant viruses in the world. The circulative propagative transmission of TSWV is carried out by at least ten different species of thrips. The most common species is Frankliniella occidentalis (western flower thrips) as it is the vector that predominantly transmits TSWV globally and in greenhouses. The rapid developmental and reproductive rate of the thrips contributes to the spread of TSWV. The amount of time it takes for insects to acquire the virus (acquisition period) and the amount of time it takes for the virus to move from the insect to the plant (inoculation) for TSWV varies depends on the vector species. For Frankliniella occidentalis, the acquisition and inoculation of TSWV can be as short as 5 minutes. However, the acquisition and inoculation periods for optimal transmission is 21.3 hours and 42.7 hours, respectively.

Transmission of TSWV can only occur when larvae stage thrips acquire TSWV. The larvae stage for thrips lasts around 1-3 days. TSWV is acquired by thrips when they feed on infected plants. Adult thrips cannot be infected with TSWV as their midgut barrier successfully prevents infection. However, thrips that have successfully become infected with TSWV in the larvae stage can transmit the virus throughout their lifetime. In order to protect their eggs, thrips insert their eggs into various types of plant tissue. Eggs can be found in the stems, leaves, or flowers of plants. Thrips hatch in 2-3 days and complete their lifecycle in 20-30 days. Adult thrips feed on the flower bud, stem and leaf parts of the plant.

Hosts and Symptoms

TSWV infects a variety of hosts, contributing to its global economic impact on crops. There are over a thousand different hosts for TSWV. The host range of TSWV includes agronomically important crops like tomatoes and tobacco. The symptoms of TSWV vary from host to host. There is also variability of symptoms within a single type of host due to the age of the plant, nutrition and the environment (especially temperature). Common symptoms include stunting, ringspots on fruit and necrosis of leaves. There are many different strains of TSWV, and differences in symptoms may also be attributed to the differences in the number of strains present.

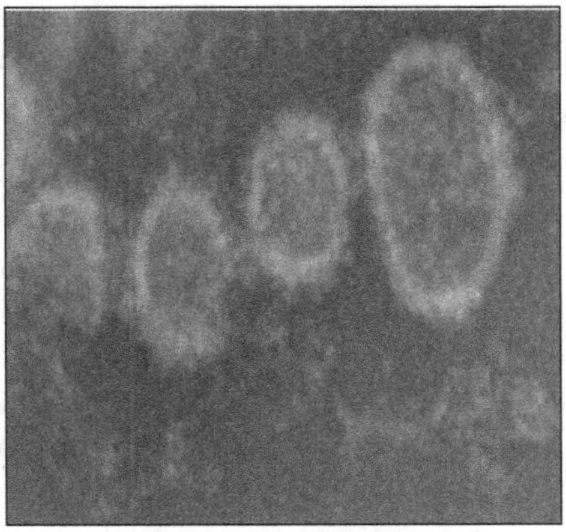

Tomato Spotted Wilt Virus.

Management

Prevention is key in managing TSWV. Once a plant becomes infected with TSWV, there are no practical ways to cure the virus infected plant. The most efficient method of containing this disease is genetic resistance. There are several different resistance genes identified in multiple crops. In some crops the resistance genes have been effective, however, in others, some strains of TSWV have been discovered to overcome the resistance gene, such as the Sw-5 resistance gene in tomato. The Sw-5 resistance gene in tomato is a dominant resistant gene. The Sw-5 gene gives resistance to the TSWV through a hypersensitive response. A hypersensitive response is when the plant cells that surround the infection undergo cell death which would then deprive the virus of the cell machinery it needs to replicate and infect the plant further. There have been several strains of TSWV detected in countries such as Australia, Spain and the United States that can overcome the Sw-5 resistant gene. However, those strains of TSWV have not been spread worldwide so the Sw-5 gene is still useful.

Other important prevention techniques include buying virus and thrips-free transplants and managing thrips populations. Introducing species that naturally prey on thrips, such as the minute pirate bugs (Orius insidiosus) and big eyed bugs (Geocoris punctipes), may help reduce transmission of TSWV. Insecticides are not an efficient way to decrease the vector population because the vectors rapidly develop resistance. Removing weeds and infected plants is a good way to prevent more infections in the greenhouse. Sanitation practices such as the destruction or removal of old crops by plowing or physical removal are often used in the field.

Potato Virus Y

Potato virus Y (PVY) is a plant pathogenic virus of the family Potyviridae, and one of the most important plant viruses affecting potato production.

PVY infection of potato plants results in a variety of symptoms depending on the viral strain. The mildest of these symptoms is production loss, but the most detrimental is 'potato tuber necrotic ringspot disease' (PTNRD). Necrotic ringspots render potatoes unmarketable and can therefore result in a significant loss of income. PVY is transmissible by aphid vectors but may also remain dormant in seed potatoes. This means that using the same line of potato for production of seed potatoes for several consecutive generations will lead to a progressive increase in viral load and subsequent loss of crop.

An increase in potato plant infection with viruses over the past few years has led to considerable losses to the South African potato industry. The increased rate of infection may be attributed to several factors. These include a marked decrease in the effectiveness and administration of chemicals used in vector control, the use of infected seed potatoes in cultivation, incorrect irrigation and farming methods as well as a lack of a sensitive, rapid and reliable method of detection. An increase in the average temperature of winters as a consequence of global warming has also led to an increase in aphid numbers, which in turn has led to an increase in viral distribution.

Potato Virus Y Hosts, Strains and Symptoms

PVY belongs to the genus Potyvirus, of which it is the type member. Potyvirus is the largest genus of plant viruses, and possibly the most destructive one in potato crops. The genus includes more than 200 species that bring about significant losses in the agricultural arena. PVY infects many economically important plant species. These include potato (Solanum tuberosum), tobacco (Nicotiana tabacum), tomato (Solanum lycopersicum) and pepper (Capsicum spp.). The level of damage to crop is determined by the strain of PVY infecting the plants, the viral load, the time at which infection occurs as well as the tolerance the host possesses toward the virus. Resistance to PVY infection by hosts is low in many cases. Infection of a potato field with PVY may ultimately result in 10-100% loss in yield.

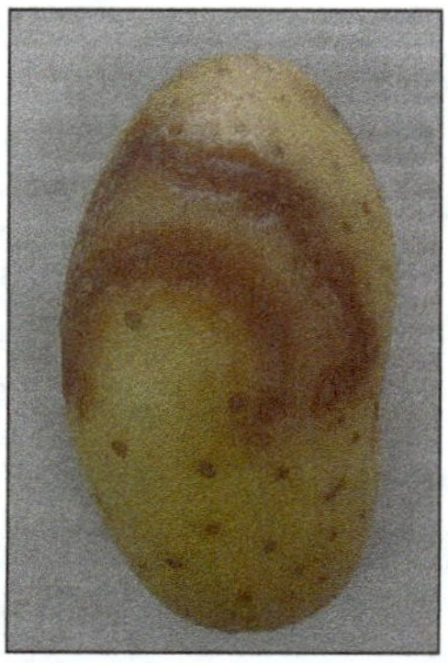

Potato illustrating necrotic ringspot disease.

It has been shown that the PVY has different isolates according to the symptoms they induce in various potato plant species. Extensive biological, serological and molecular variability of PVY isolates makes the classification of isolates as particular strains particularly difficult. Occurrence of a variety of symptoms and the emergence of the necrotic PVY^{NTN} has led to a search for more reliable classification tools than simple serological identification. Traditionally three chief strains of PVY are recognized: PVY^{C}, PVY^{N} and PVY^{O}. PVY^{C}, originally known as Potato Virus C, was the first to be recognized and was identified in the 1930s. PVY^{C} induces hypersensitive responses in a wide range of potato cultivars. These reactions include the formation of mild mosaic patterns or stipple streak. Unlike the other strains of PVY, some PVY^{C} strains are non-aphid transmissible. Previous studies by Visser et al. did not identify any of the local isolates as being PVY^{C} but it has been reported to occur to in South Africa. A second strain of PVY is PVY^{N}. Some notes on suspected variant of Solanum virus 2 (Potato virus Y). This strain was described in tobacco plants growing close to potato plants. PVY^{N} results in leaf necrosis and mild or even no damage to the tubers. The ordinary strain of PVY is denoted as PVY^{O}. Infection of a potato plant with the PVY^{O} strain results in mild tuber damage and does not cause leaf necrosis. Both PVY^{N} and PVY^{O} are aphid transmissible and occur in South Africa. In Europe these two strains have been shown to have recombined to form PVY^{NTN}. The PVY^{NTN} has been accredited with the ability to induce potato tuber necrotic ringspot disease (PTNRD). Tubers damaged by PTNRD become unmarketable and infection by PVY^{NTN} thus results in a larger economic impact than infection by the other strains.

Potato Virus Y Transmission

PVY may be transmitted to potato plants through grafting, plant sap inoculation and through aphid transmission. The most common manner of PVY infection of plant material in the field is through the aphid and although aphids on their own can directly damage potato plants it is their role as viral vectors which has the greatest economic impact. In cold climates aphids spend the winter either as wingless aphids giving birth to live young (viviparae) or as eggs. Hosts such as weeds and other crops serve as breeding grounds for these aphids and form a temporary area of colonization before the aphids migrate to the potato fields. In moderate climates, such as in South Africa, aphids are thought to reproduce asexually on weeds, other crops, indigenous plants and garden plants. This means that there are a number of aphids present year-round. The importance in effective and stringent monitoring of aphid populations is stressed in a review by Radcliffe and Ragsdale as PVY virions are introduced to potato fields almost solely by winged aphids from a virus source outside these fields. Wingless aphids have not yet been linked to the spread of PVY in potato fields.

The green peach aphid (Myzus persicae) has been found to be most effective in its role as viral vector, but others such as Aphis fabae, Aphis gossypii, Aphis nasturtii, Macrosiphum euphorbiae, Myzus (Nectarosiphon) certus, Myzus (Phorodon) humuli and Rhopalosiphum insertum are also strongly associated with viral transmission. The Agricultural Research Council-Vegetable and Ornamental Plant Institute

(ARC-VOPI) 6 of South Africa identified twenty five species of aphid able to function as PVY vectors. The efficiencies of some of these aphids to function as PVY vectors were also established and were found to vary between the different species. In South Africa, Aphis fabae, Aphis gossypii and Aphis nasturtii are the most common and efficient PVY vectors found in the field. Apart from being classed according to efficiency as vectors, aphids can also be divided into two subgroups, namely colonizing and non-colonizing species. Colonizing aphids are aphids which reproduce and establish themselves on potato plants, specifically, while non-colonizing aphids do not reproduce nor establish colonies on potato plants. Colonizing aphids are better adapted to life on potato plants and are thus generally considered as better PVY vectors than non-colonizing aphids. Noncolonizing aphids do not primarily feed on potato plants but do occasionally feed on them while searching for a more suitable host. Their lower efficiency as PVY vector is cancelled out by the sheer numbers in which they occur. Because of this, all aphids present in and around potato fields must be considered as possible vectors and their numbers carefully monitored.

Transmission of PVY by aphids occurs in a non-persistent, non-circulative manner which suggests a less intimate interaction between virion and vector than is the case of circulative virions. The fact that the virions are transmitted in a non-persistent fashion means that viral replication does not occur within the aphid vector and that, unless the aphid feeds on infected plants, it loses its ability to infect plants after two to three feedings. The virions attach to the aphid stylet in a matter of seconds and may remain infectious for four to seventeen hours. The distance over which the virions can be transmitted is limited due to the short period for which they remain infectious. Although the short life span outside plants inhibits long-distance viral transmission, it does not reduce the transmission efficiency bestowed by the quick rate of viral acquisition and inoculation within a field.

Upon entrance into the plant cell, the virus coat protein disassembles and releases its RNA genome. The viral RNA serves as mRNA, and although little is known about the translation thereof, it is believed that the 5' non-coding region functions as an enhancer of translation. The translated mRNA results in a polyprotein which is processed into mature proteins. Each polyprotein is then cleaved into ten different proteins which are believed to be multifunctional. These proteins, along with host proteins, assemble to form a replication complex. This complex performs negative-strand RNA synthesis, using the positive strand of viral RNA as a template. Once the additional RNA copies have been produced, they code for the synthesis of various proteins, as well as coat proteins. These coat proteins will now enclose the newly formed genomes to give rise to new virions. It has been suggested that enclosure of the newly formed virions is initiated by the interaction of the coat proteins with the 5'terminus and that the coat protein is built up towards the 3'terminus. The entire process of viral replication occurs within the endoplasmic reticulum. These newly synthesized viral particles are subsequently transported through the plasmodesmata to adjacent plant cells via several assisting potyvirus proteins. Distribution of viruses within the plant occurs according to the source-sink relationship between maturing and growing tissues. Virus concentration throughout the plant is high and this greatly increases the chance of uptake by aphids. Infection of plants by potyviruses can be varied in the symptoms shown. Infection can include veinal necrosis, mosaic symptoms as well as leaf malformation. Infected plants that do not show symptoms may have infected canopies and will yield lower quality products than their healthy counterparts.

Potato – PVY^{NTN} Interaction

Since PVY^{NTN} causes great loss in potato production, the research of potato – potato virus Y^{NTN} interaction is important. Sensitive potato cultivars respond to PVY^{NTN} inoculation with development of typical symptoms. On inoculated leaves 5 – 7 days after inoculation chlorotic and necrotic ringspots develop. As the virus spreads through the plant the systemic symptoms develop on uninoculated leaves. 10 days after inoculation wrinkles and mosaic chlorosis appear, leading to a palm tree appearance (leaf drop).

The viral defense mechanisms of plants will primarily try to restrict the movement of the virus. In failing this, it may attempt to induce cell death in infected tissue, thereby preventing the spread of virions. Although the precise mechanism of disease induction by potyviruses in plants is unknown, it is known that these viruses cause a significant shutdown of host gene expression during viral replication.

Physiological changes in potato plants as a response to PVY^{NTN} infection were intensively studied. At early stages of infection, meaning first 12 hours, photosynthesis related genes, genes involved in perception, signalling and defence response were shown to be differentially expressed. 24 h after inoculation the amount of salicylic acid increased.

A disruption in gene expression disrupts the normal cellular function of cells which could be the cause of the physical symptoms that the plant demonstrates. At the time of symptoms development, research on interaction between susceptible potato cultivar and PVY^{NTN} showed changes in cytokinin level. In inoculated leaves showing symptoms modifications in chloroplast structure and size, lower chlorophyll levels and differential activity of soluble and ionically-bound peroxidases were detected.

At later stages of PVY^{NTN} infection total protein concentration increased in sensitive potato cultivar while no such pronounced changes were observed in tolerant and moderately tolerant potato cultivars. Gene expression studies revealed changes in expression of genes for heat-shock proteins, catalase, β-1,3-glucanase and genes involved in photosynthesis.

Molecular Description of Potato Virus Y

Potyvirus virions consist of non-enveloped filamentous structures that are 680 – 900 nm in length and 11 to 15 nm in width. Morphologically the potyvirus capsid consists of approximately 2 000 copies of coat protein (CP).

The capsid encapsulates a single strand of positive sense RNA which is in the order of 10 kb in length and has a non-translated 5'-terminal region (5'-NTR) as well as a 3'-poly-A tail. The positive sense genome contains a single extended open reading frame and acts directly as mRNA. The 144 nucleotide 5'-NTR is particularly rich in adenine residues and has very few guanine residues. Rather than a conventional cap structure, the 5'NTR is associated with a Viral genome linked protein (VPg) which is said to act as an enhancer of transcription.

The 5'-leader sequence has an internal ribosome entry site (IRES) and cap-independent translation regulatory elements (CIREs). The IRES directs cap-independent translation through a mechanism similar to that used by eukaryotes. The extended open reading frame encodes for a 350 kDa polyprotein. This polyprotein is proteolytically processed by viral proteases (NIa, HC-Pro and P1) and undergoes co- and post-translational cleavage to yield several multi-functional proteins. These include the following: P1 (P1 Protein), HCPro (Helper Component Proteinase), P3 (P3 Protein), 6K1 (6-kDa Protein 1), CI (Cylindrical Inclusion), 6K2 (6-kDa Protein 2), VPg (Viral Protein genome-linked), NIaPro (Nuclear Inclusion Protein a, Proteinase domain), NIb (Nuclear Inclusion Protein b) and the CP (Coat Protein).

Diagnostic Techniques for Detection of Potato Virus Y

ELISA

In the past, crops were inspected visually to determine whether or not they were disease free. Visual inspection was also used as a basis for seed certification. Determination of viral status through visual inspection is incredibly difficult as the symptoms may be masked or the infection latent. As a result, post season tests and inspections were introduced. These tests involved the cultivation of previously harvested material in greenhouses. The resulting plants were inspected for a more accurate estimate of viral status. Although this method of screening did offer some degree of monitoring of viral presence it was subjective and highly ineffective. Enzyme-linked immunosorbent assay (ELISA) screening of crops and seed potatoes replaced visual inspection in the early 1970s. The use of ELISA offered routine diagnostic laboratories a quick, effective and sensitive method of screening for a wide range of potato plant viruses.

Detection of pathogens using ELISA relies on the interaction between the antigen and specific antibodies and has become a popular and cost-effective means of routine detection. In an ELISA the solid phase can be coated with the sample of interest containing the antigen. The efficiency to which the antigen binds to the solid phase is dependent on temperature, length of exposure as well as concentration. Solid phases used include nitrocellulose membranes, paper, glass, agarose and polystyrene or polyvinylchloride microtiter plates. Microtiter plates are the most widely used solid phase because they are easy to handle, allow for automation and for analysis using microtiter plate readers. A drawback of these plates is that they are highly absorptive and this increases the incidence of non-specific binding of components used in the ELISA. Non-specific binding to the plates is reduced through the use of buffers containing proteins such as casein and non-ionic detergents such as Tween 20. After coating, excess sample is removed and the plate typically treated with a 1% casein containing solution.

Subsequent to this the solid phase is treated with antibodies raised against the antigen of interest. After each incubation step the plate is washed with Tween 20 containing PBS. These washing steps are aimed to wash away any non-specifically bound components. Nonspecifically bound components are less strongly bound than the specific bound ones. Detection is achieved either through the addition of an enzyme-coupled antibody or the addition and detection of a biotinylated antibody. In a system using an enzyme-coupled antibody the subsequent addition of an appropriate substrate results in the formation of a colour proportional to the amount of antigen. Alternatively the plate can be coated with antibody followed by incubation with the sample that is to be detected. This, in turn, can be detected as described above and is then referred to as the double antibody sandwich (DAS) ELISA. Both of these systems, however, have a disadvantage in that coupling of the enzyme to the antibody may result in steric hindrance which in turn may result in a loss in function of the antibody and the enzyme. This may be overcome through the use of a biotin-avidin or biotin-streptavidin bridge. In this type of system biotin is coupled to the antibody. The biotin molecule has no influence on the working of the antibodies and is easily detectedusing avidin or streptavidin conjugated to a suitable enzyme. Streptavidin has an extremely high affinity for biotin which results in even a higher degree of specificity than a system in which the enzyme is coupled directly the antigen. To establish whether or not the antigen is present, a substrate specific for the enzyme used is added. The enzyme then converts the substrate to a coloured product and the colour intensity can be correlated to the amount of antibodies bound and thus the amount of antigen present. A DAS-ELISA has the advantage that it can increase the specificity of the ELISA and reduce the occurrence of non-specific binding. As a result, the DAS-ELISA principle is commonly employed in ELISA's for the detection of plant pathogens in plant sap without prior purification of the pathogen.

The ELISA is considered to be a safe, inexpensive and rapid method for detection of plant viruses. The inexpensive nature and relative simplicity thereof allows for it to be used as a workhorse within the agricultural sector and is used to screen thousands of samples per year. Unfortunately ELISAs are not completely failsafe. Virus levels within potato tubers, which are screened by ELISA for use as seed potatoes, are normally low while the tubers are dormant. ELISA detection of viruses in these potatoes is difficult and absorbance values may fall below the set cut-off value. For this reason, seed tuber screening is performed on sprouting rather than dormant tubers. Although this results in more reliable readings than direct tuber testing, it does delay the certification of seed potatoes. Another disadvantage of an immuno-based detection method is that changes at the gene level may have an influence on the immunogenicity of the antigen to be detected. In terms of potato plant viruses, mutations within the CP gene may cause the CP to undergo conformational changes rendering antibodies produced against the previously present virus less effective.

RT-PCR

Reverse transcriptase polymerase chain reaction (RT-PCR) has become a powerful and effective method for detection of potato plant viruses within potato plant material and even dormant potatoes. Only a minute piece of plant material is required for analysis using RT-PCR. Considering the protocol described within this thesis, 0.1 g of plant material is enough for 14 500 separate reactions. During a RT-PCR specific target RNA sequences are amplified exponentially into DNA copies. For this to occur, however, the RNA of the virus must first be transcribed to DNA by means of a reverse transcriptase polymerase. This polymerase synthesizes a DNA strand using the RNA as template. This results in a DNA/RNA complex. For synthesis of a DNA strand from the RNA template only the reverse primer is required since the RNA is a single strand arranged from 5' to 3'. Subsequently, the newly synthesized DNA strand is used as a template for traditional PCR.

Different types of reverse transcriptase polymerases are available to suit different needs and reaction conditions. Reverse transcriptase enzymes commonly used include AMV RT, SuperScript III, ImProm-II, Omniscript, Sensiscript and Tth RT. At the end of the RT step the polymerase enzyme is heatactivated. It could also be that the reverse transcriptase polymerase and DNA polymerase is one and the same enzyme and that the enzyme only requires a DNA polymerase activation step after the RT step. An example of such an enzyme is Tth polymerase. This enzyme has both RNA-dependent reverse transcriptase and DNA-dependent polymerase activity. However, the active center of the DNA polymerase is covered by dedicated oligonucleotides, called aptamers. At temperatures below the optimal reaction temperature of the DNA-dependent polymerase component of Tth remains covered by the aptamers. At these temperatures the Tth enzyme only synthesizes a DNA copy of the RNA template. Once the reaction temperature is raised to 95 °C, the aptamers are removed and the DNA-dependent polymerase component will start to amplify the target sequence.

PCR amplification of the DNA target occurs in three steps: denaturation, annealing and extension. Each of these steps occur at a specific temperature for a fixed period of time. Denaturation is normally allowed to occur between 90 and 95 °C and leads to the dissociation of DNA strands. After this the reaction is cooled to between 40 and 70 °C to allow the primers to associate with their respective target sequences. This step is known as the annealing step and is primer specific. The temperature at which the primers anneal is critical. Too high temperatures would not allow the primers to associate with the DNA, resulting in no or poor amplification. Too low annealing temperature would ultimately lead to non-specific binding of the primers and non-specific amplification. Primers bound to the regions flanking the target DNA provide 3'-hydroxyl groups for DNA polymerase catalyzed extension. The most commonly used DNA polymerase is Taq, a thermo-stable enzyme isolated from the thermophilic bacterium, Thermus aquaticus. The DNA polymerase synthesizes new DNA strands along the template strands, using the primers as starting points. During the extension step the strands are amplified beyond the target DNA. This means that each newly synthesized strand of DNA will have a region complementary to a primer.

There is an exponential increase in the amount of DNA produced as the three above mentioned steps are repeated in a cyclic fashion. In a traditional PCR these steps might be repeated 20 to 55 times. A problem, however, with PCR amplification is that the temperature required for DNA strand dissociation also results in DNA polymerase denaturation. This is partially overcome by the bioengineering of polymerases which are more thermal stable and have longer half-lives.

Even though RT-PCR is technically more difficult to perform and more expensive than ELISA, it has the ability to allow for the detection of low viral loads. RT-PCR is considered to be 102 to 105 fold more sensitive than traditional ELISA. RT-PCR also allows for the detection of several viral targets in the same reaction through the use of several primer combinations. This is called multiplexing. Although multiplexing is technically more demanding than a traditional simplex reaction it allows for a higher throughput in that a single sample can be tested for several viral strains in a single reaction. Primers used for multiplexing are chosen in such a manner that they result in amplicons of various sizes. This allows for post RT-PCR analysis using gel electrophoresis. Although RT-PCR saves time, allows for multiplexing and is more sensitive than ELISA, the reagents and instrumentation needed are expensive and require a higher level of technical expertise. Also, end product analysis using gel electrophoresis is laborious, relatively more expensive, time consuming and does not lend itself to automation. For these reasons the use of RT-PCR for routine screening is not feasible and has not replaced ELISA. It does, however, provide the industry with the opportunity to screen borderline cases, especially in the case of seed potato certification.

Quantitative PCR

In most traditional PCRs the resulting products are analyzed after the PCR has been completed. This is called end-point analysis and is normally qualitative of nature rather than being quantitative. For this sort of analysis, products are mostly analyzed on an agarose gel and visualized using ethidium bromide as a fluorescent dye. Direct correlation between signal strength and initial sample concentration is not possible using end-point analysis since PCR efficiency decreases as the reaction nears the plateau phase. Quantitative PCR, however, offers an accurate and rapid alternative to traditional PCR. Quantitative PCR offers the researcher the opportunity to amplify and analyze the product in a single tube using fluorescent dyes. This is known as homogeneous PCR. During a quantitative PCR the increase in fluorescence is correlated with the increase in product. Through the use of different specific, dyes quantitative PCR can be used to distinguish between different strains of a virus and even to detect point mutations. The major advantage of quantitative PCR is that analysis of resulting products using gel electrophoresis is not required. This means that quanti-tative PCR can be implemented as a high-throughput technique for sample screening.

Quantitative PCR has been described for detection and discrimination of PVY^{O} and PVY^{N} isolates and for reliable discrimination between PVY^{NTN} and PVY^{N} isolates.

Potato Virus X

Potato virus X (PVX) is a plant pathogenic virus of the family Alphaflexiviridae and the order Tymovirales. It is the type species of the genus Potexvirus.

PVX is found mainly in potatoes and is only transmitted mechanically. There are no insect or fungal vectors for this virus. This virus causes mild or no symptoms in most potato varieties, but when Potato virus Y is present, synergy between these two virus es causes severe symptoms in potatoes. The virion has helical symmetry and a deeply grooved, highly hydrated surface and is made of a single-stranded positive-sense RNA genome of approximately 6.4 kb. This is wrapped in approximately 1300 units of a single coat protein (CP) type, with 8.9 CP units per helix turn. The genome is capped at the 5′-end and poly-adenylated at the 3′-terminus. It contains five open reading frames (ORFs) encoding five proteins: the RNA-dependent RNA Polymerase (RdRP), the movement proteins encoded by three overlapping ORFs that form the Triple Gene Block module (TGBp1, TGBp2, and TGBp3), and the CP (coat protein).

Virus indexing and limited generation production of potato, which starts from dis ease-free tissue culture plantlets, has nearly eliminated this virus from many countries' potato supply.

Human Virome

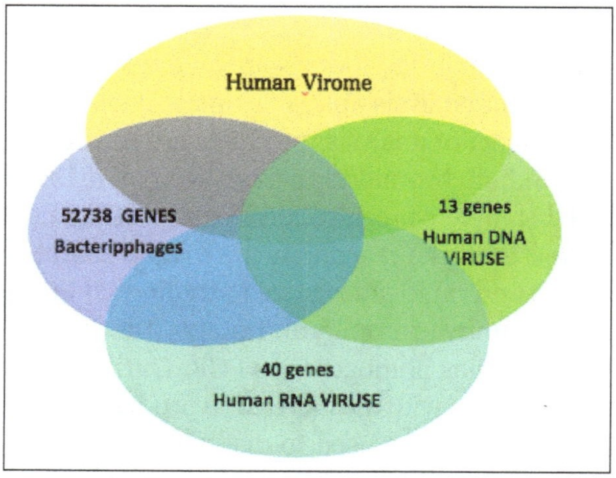

The diversity of human viral sequences.

The human virome is the total collection of viruses in and on the human body. Viruses in the human body may infect both human cells and other microbes such as bacteria (as with bacteriophages). Some viruses cause disease, while others may be asymptomatic. Certain viruses are also integrated into the human genome as proviruses or endogenous viral elements.

Viruses evolve rapidly and hence the human virome changes constantly. Every human being has a unique virome with a unique balance of species. Lifestyle, age, geographic location, and even the season of the year can affect an individual's exposure to viruses, and one's susceptibility to any disease that might be caused by those viruses is also affected by pre-existing immunity and both viral and human genetics.

The human virome is far from being completely explored and new viruses are discovered frequently. Unlike the roughly 40 trillion bacteria in a typical human microbiome, an estimate of the number of viral particles in a healthy adult human is not yet available, although virions generally outnumber individual bacteria 10:1 in nature. Studying the virome is thought to provide an understanding of microbes in general and how they affect human health and disease.

Methods and Tools

Multiple methods are available for the isolation and study of human viruses:

- Deep sequencing is a rapid DNA sequencing technique that is useful for characterizing virome richness, stability, gene function and the association with disease phenotypes. This technology creates large amounts of sequence information and is capable of detecting rare components of a microbial community. Current methods combining the removal of human and bacterial DNA from samples, large scale sequencing, and bioinformatics are very efficient in the identification of unknown viruses. Unlike other discovery methods, viruses do not need to be grown in cell cultures. Without any prior knowledge of genome sequence or growth methods, novel viruses can be discovered. Therefore, deep sequencing is well suited for rapid identification of an unknown or unexpected viruses involved in a disease outbreak or associated with conditions not thought to be caused by viruses. Deep sequencing also allows for large scale screenings with minimal hands on effort. A systematic exploration of the viruses that infect humans (the human virome) is important and feasible with these methods.
- Polymerase chain reaction is a tool to amplify and detect specific DNA sequences. It can be used to help characterize the virome, but it is limited by the need for at least partial DNA sequence information.
- The human metagenome includes all organisms that live on or in us. Viruses contribute to the metagenome and establish chronic infection that infest

chromosomes; this method will formulate new estimate of the number of genes that confer susceptibility to a given virus and specify alleles for some viruses.

- Large scale antibody studies with ELISA using donated blood could help to determine human exposure to particular viruses in different geographic regions.

Diversity of Human Viruses

The human virome is not stable and may change over time. In fact, new viruses are discovered constantly. With an increasing number of known viruses, diagnosis and treatment of novel viral-associated conditions will become easier as well. Studying the virome could help improve drug development and limit antibiotic usage.

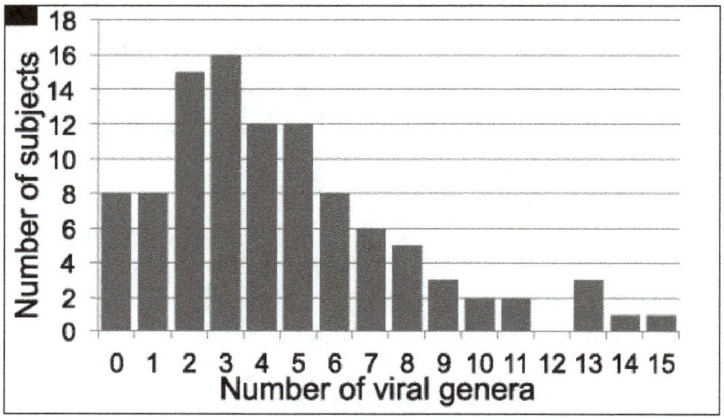

The human virome in healthy, asymptomatic adults. The histogram shows the number of individuals (y-axis) who were positive for a given number of different viral genera (x-axis).

One of the first studies that used high-throughput DNA sequencing to describe the diversity of eukaryotic dsDNA viruses in normal individuals included 706 samples from 102 subjects. This study detected an average of 5.5 viral genera in each individual and these viruses included herpesviruses, papillomaviruses, polyomaviruses, adenoviruses, anelloviruses, parvoviruses, and circoviruses.

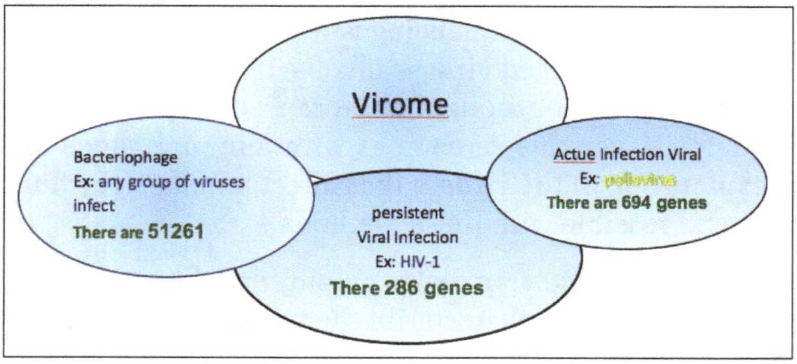

The human virome in healthy, asymptomatic adults. The viral genera (x-axis) detected in each subject (y-axis) are represented by black bars. The virome of each individual is viewed by looking at the black bars in a given row.

Each individual had a distinct viral profile, demonstrating the high interpersonal diversity of the virome. One to 15 viral genera (average 5.5) were detected in 92% of the 102 individuals sampled. Figure illustrates the viromes of the 102 individuals defined by sampling up to five major body habitats, showing that a broad range of viruses was detected in healthy people.

The 102 individuals carried seven distinct families of human DNA viruses. Sequences were detected predominantly in the nose and skin, similarity to 17 papillomavirus genera. Roseoloviruses, predominantly HHV-7 and to a lesser extent HHV-6, were present among 98% of the individuals who provided mouth samples.

In addition, the same viruses were prevalent in multiple body habitats within individuals. For instance, the beta- and gamma-papillomaviruses were the viruses most commonly found in the skin and the nose, which may reflect proximity and similarities in microenvironments that support infection with these viruses.

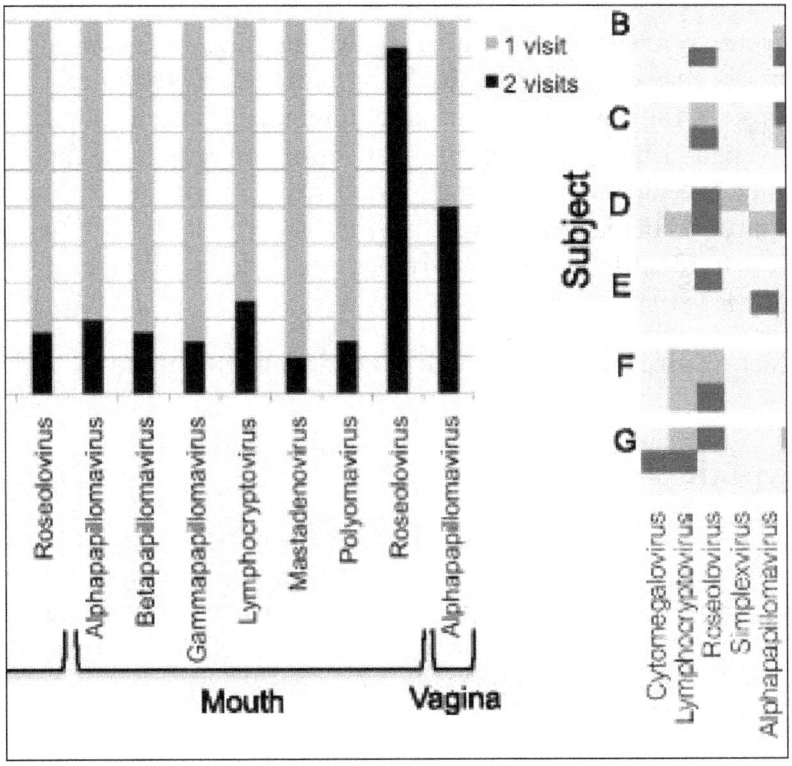

The human virome in five body habitats. (A) All of the viruses detected in the five body habitats. Each virus is represented by a colored bar and labeled on the y-axis on the right side. The relative height of the bar reflects the percentage of subjects sampled at each body site in whom the virus was detected. In this panel, the bar representing roseoloviruses in the oral samples reflects the maximum bar height, because 98% of the individuals who were sampled in the mouth harbored roseoloviruses. (B) This panel shows papillomaviruses included in the category 'Other papillomaviruses'. The largest bar height shown represents the unclassified papillomaviruses found in skin samples from 65% of subjects.

Human Blood Virome

Whole-genome sequencing data of blood from 8,240 individuals without any clear infectious disease revealed 94 different viruses in 42% of the study participants. The sequences included 19 human DNA viruses, proviruses and RNA viruses (herpesviruses, anelloviruses, papillomaviruses, three polyomaviruses, adenovirus, HIV, HTLV, hepatitis B, hepatitis C, parvovirus B19, and influenza virus). Of possible relevance to transfusion medicine, this study identified Merkel cell polyomavirus in 49 individuals, papillomavirus in blood of 13 individuals, parvovirus B19 in 6 individuals, and the presence of herpesvirus 8 in 3 individuals.

Impact on Human Health

The human virome is a part of our bodies and will not always cause harm. Many latent and asymptomatic viruses are present in the human body all the time. Viruses infect all life forms; therefore the bacterial, plant, and animal cells and material in our gut also carry viruses. When viruses cause harm by infecting the cells in the body, a symptomatic disease may develop. Contrary to common belief, harmful viruses may be in the minority compared to benign viruses in the human body. It is much harder to identify viruses than it is to identify bacteria, therefore our understanding of benign viruses in the human body is very rudimentary.

The health effects of viruses on an individual are dependent on the individual's immune system.

Hygiene Hypothesis

Recent research has linked the emerging idea of the hygiene hypothesis to viruses. This hypothesis attempts to explain and justify some of the high incidences of diseases such as asthma and eczema in the Western world to Western society's overuse of antibiotic and antiviral agents. This overuse potentially disrupts not only the bacteria of the gut but also the viruses that have long lived in harmony with the human body and now play a role in regulation of human health. This hypothesis generally refers to microorganisms but is now being extended to include airborne viruses and common viral infections of childhood that are becoming increasingly less common.

Effect of Skin Location

Diverse viruses colonize the human skin and differ by skin site. These skin virome includes human viruses (i.e. human papillomavirus) and bacteriophages (bacterial viruses) that infect commensal skin bacteria such as Staphylococci. Virus communities differ by moisture levels and degree of protection from the external environment.

Effect of Diet

Many studies have demonstrated that the bacteria and viruses in the human gut (the gut microbiome) can be altered by changes in diet. One study that focused on bacterial viruses, called bacteriophages, in the gut found a significant relationship between diet and the type of bacteriophages present. This was done by comparing the distance between bacteriophage gut communities in individuals both before and after they started a controlled diet. The results were that the distance between the bacteriophage gut communities of individuals on the same diet was significantly smaller at the end of their dietary treatment than it was at the start, while there was no increase in community similarity for individuals on different diets over time.

Human Viruses in Water

Viruses are a major cause of human waterborne and water-related diseases. Waterborne diseases are caused by water that is contaminated by human and animal urine and feces that contain pathogenic microorganisms. A subject can get infected through contact with or consumption of the contaminated water. Viruses affect all living organisms from single cellular plants, bacteria and animal to the highest forms of plants and animals including human beings. Viruses can interact with their host through several methods and mechanisms; some viruses can be host specific (HIV) and some can be less host specific (influenza) viruses. Different viruses can have different routes of transmission; for example, HIV is directly transferred by contaminated body fluids from an infected host into the tissue or bloodstream of a new host while influenza is airborne and transmitted through inhalation of contaminated air containing viral particles by a new host. Enteric viruses primarily infect the intestinal tract through ingestion of food and water contaminated with viruses of fecal origin. Some viruses can be transmitted through all three routes of transmission.

Water virology started about half a century ago when scientists attempted to detect the polio virus in water samples. Since then, other pathogenic viruses that are responsible for gastroenteritis, hepatitis, and many other virus strains have replaced enteroviruses as the main aim for detection in the water environment.

Water virology was born after a large hepatitis outbreak transmitted through water was confirmed in New Delhi between December 1955 and January 1956.

Viruses can cause massive human mortality. The smallpox virus killed an estimated 10 to 15 million human beings per year until 1967. Smallpox was finally eliminated in 1977 by extinction of the virus through vaccination, and the impact of viruses such as influenza, poliomyelitis and measles are mainly controlled by vaccination.

Despite advances in vaccination and prevention of viral diseases, it is estimated that in the 1980s a child died approximately every six seconds from diarrhea confirmed by WHO. Many cases of hepatitis A and E, both of which are enteric viruses, are typically transmitted by food and water. Extreme examples include the outbreak of 300,000 cases of hepatitis A and 25,000 cases of gastroenteritis in 1988 in Shanghai caused by shellfish harvested from a sewage polluted estuary. In 1991, an outbreak of 79,000 cases of hepatitis E in Kanpur was ascribed to drinking polluted water.

A more recent outbreak of Hepatitis E in South Sudan killed 88 people. Medecins Sans Frontieres (MSF) said it had treated almost 4,000 patients since the outbreak was identified in South Sudan in July 2012. In this outbreak, Hepatitis E, which causes liver infections, and was thought to be spread by drinking water contaminated with feces.

Sewage contaminated water contains many viruses, over one hundred species are reported and can lead to diseases that affect human beings. For example, hepatitis, gastroenteritis, meningitis, fever, rash, and conjunctivitis can all be spread through contaminated water. More viruses are being discovered in water because of new detection and characterization methods, although only some of these viruses are human pathogens.

Family	Genus	Species	Common Name	Disease Caused
Adenovirus	Mastadenovirus	Human mastadenovirus A through G	Adenovirus	Adenovirus infection, pharyngitis, conjunctivitis, fever
Astrovirus	Mamastrovirus	Human astrovirus	Astrovirus	Gastroenteritis, diarrhea
Calicivirus	Norovirus	Norwalk virus	Norovirus, winter vomiting bug	Gastroenteritis, fever
Coronaviridae	Coronavirinae	SARS coronavirus	SARS-CoV	SARS, gastroenteritis, respiratory disease
Coronaviridae	Torovirus	Human torovirus	Torovirus	Gastroenteritis
Hepeviridae Orthohepevirus		Orthohepevirus A	Hepatitis E virus, HEV	Hepatitis E
Picornavirus	Enterovirus	Enterovirus A	Coxsackie A virus	Hand, foot, and mouth disease, paralysis, meningitis, fever, respiratory disease, myocarditis, heart anomalies
Picornavirus	Enterovirus	Enterovirus B	Echovirus	Meningitis, fever, respiratory disease, rush, gastroenteritis
Picornavirus	Enterovirus	Enterovirus C	Poliovirus	Polio
Picornavirus	Hepatovirus	Hepatovirus A	Hepatitis A virus, HAV	Hepatitis A
Polyomaviridae	Polyomavirus	JC virus	JC virus	Progressive multifocal leukoencephalopathy
Reovirus	Rotavirus	Rotavirus A, B, and C	Rotavirus	Gastroenteritis

Virus Survival in Water

Viruses need a suitable environment to survive in. There are many characteristics that control the survival of viruses in water such as temperature, light, pH, salinity, organic matter, suspended solids or sediments, and air–water interfaces.

Temperature

Temperature has the highest effect on virus's survival in water since lower temperatures are the key to longer virus survival. The rate of protein, nucleic acid denaturation and chemical reactions that destroy the viral capsid are increased at higher temperatures, thus viruses will survive best at low temperatures. Hepatitis A, adenoviruses and parvoviruses have the highest survival rate in low temperatures amongst enteric viruses.

Light

Ultraviolet light (UV) is the light in sunlight and can inactivate viruses by causing cross-linking of the nucleotides in the viral genome. Many viruses in water are exterminated in the presence of sunlight. The combination of higher temperatures and more UV in the summer time corresponds to shorter viral survival in summer compared to winter. Double stranded DNA viruses like adenoviruses are more resistant to UV light inactivation than enteroviruses because they can use their host cell to repair the damage caused by the UV light.

Visible light can also affect virus survival by a process called photodynamic inactivation but the length and intensity of the light exposure can change the inactivation rate.

pH

The pH of most natural water is between 5-9. Enteric viruses are stable in these conditions. On the other hand, many enteric viruses are more stable at pH 3-5 than at pH 9 and 12. Enteroviruses can survive at pH 11-11.5 and 1-2, but for only short periods. Adenoviruses and rotaviruses are delicate to a pH of 10 or greater and leads to inactivation.

Virus Removal from Water

Water that is intended for drinking should go through some treatment to reduce pathogenic viral and bacterial concentrations. As the density of the human population has increased the incidence of sewage contamination of water has increased as well, thus the risk to humans from pathogenic viruses will increase if precautions are not taken.

Scientific studies suggest that the most common viruses found are caliciviruses, astroviruses and enteric viruses. Laboratories are still looking for improved methods to detect these pathogenic viruses. Reducing the amount of viruses in drinking water is accomplished by various treatments that are typically part of drinking water treatment systems in developed countries.

Water purification of surface water (water from lakes, rivers, or reservoirs) typically utilizes four treatment stages: coagulation and flocculation, sedimentation, filtration, and disinfection. The first three stages remove mainly dirt and larger particles, although filtration does reduce the number of viruses and bacteria in the water the number of pathogens present after filtration is still considered too high for drinking water. Purification of water from underground aquifers, called ground water, may skip some of these steps as ground water tends to have fewer contaminants than surface water. The last step, disinfection, is primarily responsible for the reduction of pathogenic viruses to safe levels in all drinking water sources. The most common disinfectants used are chlorine and chloramine. Ozone and UV light can also be used to treat large volumes of water to remove pathogens.

Prevention of Water Viruses

The quality of drinking water is ensured through a framework of water safety plans that ensures the safe disposal of human waste so that drinking water supplies are not contaminated. Improving the water supply, sanitation, hygiene and management of our water resources could prevent ten percent of total global disease.

Half of the hospital beds occupied in the world are related to the lack of safe drinking water. Unsafe water leads to the 88% of the global cases of diarrhea and 90% of the deaths of diarreaheal diseases in children under five years old. Most of these deaths occur in developing countries due to poverty and the high cost of safe water.

Approximately 1.1 billion people do not have access to improved water and 2.4 billion people do not have access to sanitation facilities. This situation leads to 2 million preventable deaths each year.

Human Immunodeficiency Viruses

The human immunodeficiency viruses (HIV) are two species of Lentivirus (a subgroup of retrovirus) that infect humans. Over time they cause acquired immunodeficiency syndrome (AIDS), a condition in which progressive failure of the immune system allows life-threatening opportunistic infections and cancers to thrive. Without treatment, average survival time after infection with HIV is estimated to be 9 to 11 years, depending on the HIV subtype. In most cases, HIV is a sexually transmitted infection and occurs by contact with or transfer of blood, pre-ejaculate, semen, and vaginal fluids. Research has shown (for both same-sex and opposite-sex couples) that HIV is untransmissable through condomless sexual intercourse if the HIV-positive partner has a consistently undetectable viral load. Non-sexual transmission can occur from an infected mother to her infant during pregnancy, during childbirth by exposure to her blood or vaginal fluid, and through breast milk. Within these bodily fluids, HIV is present as both free virus particles and virus within infected immune cells.

HIV infects vital cells in the human immune system, such as helper T cells (specifically $CD4^+$ T cells), macrophages, and dendritic cells. HIV infection leads to low levels of $CD4^+$ T cells through a number of mechanisms, including pyroptosis of abortively infected T cells, apoptosis of uninfected bystander cells, direct viral killing of infected cells, and killing of infected $CD4^+$ T cells by $CD8^+$ cytotoxic lymphocytes that recognize infected cells. When $CD4^+$ T cell numbers decline below a critical level, cell-mediated immunity is lost, and the body becomes progressively more susceptible to opportunistic infections, leading to the development of AIDS.

Virology

Classification

Comparison of HIV species				
Species	Virulence	Infectivity	Prevalence	Inferred origin
HIV-1	High	High	Global	Common chimpanzee
HIV-2	Lower	Low	West Africa	Sooty mangabey

HIV is a member of the genus Lentivirus, part of the family Retroviridae. Lentiviruses have many morphologies and biological properties in common. Many species are infected by lentiviruses, which are characteristically responsible for long-duration illnesses with a long incubation period. Lentiviruses are transmitted as single-stranded, positive-sense, enveloped RNA viruses. Upon entry into the target cell, the viral RNA genome is converted (reverse transcribed) into double-stranded DNA by a virally encoded enzyme, reverse transcriptase, that is transported along with the viral genome in the virus particle. The resulting viral DNA is then imported into the cell nucleus and integrated into the cellular DNA by a virally encoded enzyme, integrase, and host co-factors. Once integrated, the virus may become latent, allowing the virus and its host cell to avoid detection by the immune system, for an indeterminate amount of time. The HIV virus can remain dormant in the human body for up to ten years after primary infection; during this period the virus does not cause symptoms. Alternatively, the integrated viral DNA may be transcribed, producing new RNA genomes and viral proteins, using host cell resources, that are packaged and released from the cell as new virus particles that will begin the replication cycle anew.

Two types of HIV have been characterized: HIV-1 and HIV-2. HIV-1 is the virus that was initially discovered and termed both lymphadenopathy associated virus (LAV) and human T-lymphotropic virus 3 (HTLV-III). HIV-1 is more virulent and more infective than HIV-2, and is the cause of the majority of HIV infections globally. The lower infectivity of HIV-2, compared to HIV-1, implies that fewer of those exposed to HIV-2 will be infected per exposure. Due to its relatively poor capacity for transmission, HIV-2 is largely confined to West Africa.

Structure and Genome

HIV is different in structure from other retroviruses. It is roughly spherical with a diameter of about 120 nm, around 60 times smaller than a red blood cell. It is composed of two copies of positive-sense single-stranded RNA that codes for the virus's nine genes enclosed by a conical capsid composed of 2,000 copies of the viral protein p24. The single-stranded RNA is tightly bound to nucleocapsid proteins, p7, and enzymes needed for the development of the virion such as reverse transcriptase, proteases, ribonuclease and integrase. A matrix composed of the viral protein p17 surrounds the capsid ensuring the integrity of the virion particle.

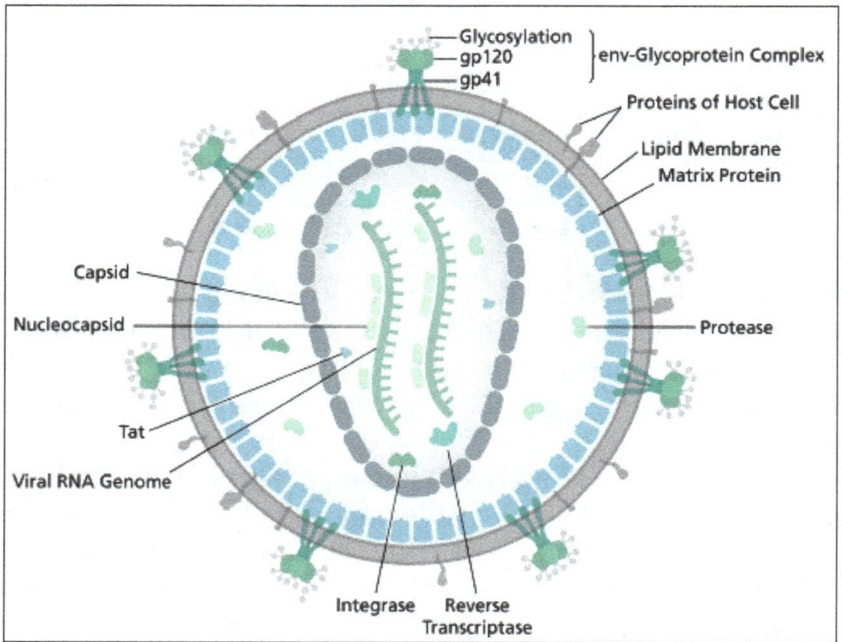

Diagram of the HIV virion.

This is, in turn, surrounded by the viral envelope, that is composed of the lipid bilayer taken from the membrane of a human host cell when the newly formed virus particle buds from the cell. The viral envelope contains proteins from the host cell and relatively few copies of the HIV Envelope protein, which consists of a cap made of three molecules known as glycoprotein (gp) 120, and a stem consisting of three gp41 molecules that anchor the structure into the viral envelope. The Envelope protein, encoded by the HIV env gene, allows the virus to attach to target cells and fuse the viral envelope with the target cell's membrane releasing the viral contents into the cell and initiating the infectious cycle.

As the sole viral protein on the surface of the virus, the Envelope protein is a major target for HIV vaccine efforts. Over half of the mass of the trimeric envelope spike is

N-linked glycans. The density is high as the glycans shield the underlying viral protein from neutralisation by antibodies. This is one of the most densely glycosylated molecules known and the density is sufficiently high to prevent the normal maturation process of glycans during biogenesis in the endoplasmic and Golgi apparatus. The majority of the glycans are therefore stalled as immature 'high-mannose' glycans not normally present on human glycoproteins that are secreted or present on a cell surface. The unusual processing and high density means that almost all broadly neutralising antibodies that have so far been identified (from a subset of patients that have been infected for many months to years) bind to, or are adapted to cope with, these envelope glycans.

The molecular structure of the viral spike has now been determined by X-ray crystallography and cryogenic electron microscopy. These advances in structural biology were made possible due to the development of stable recombinant forms of the viral spike by the introduction of an intersubunit disulphide bond and an isoleucine to proline mutation (radical replacement of an amino acid) in gp41. The so-called SOSIP trimers not only reproduce the antigenic properties of the native viral spike, but also display the same degree of immature glycans as presented on the native virus. Recombinant trimeric viral spikes are promising vaccine candidates as they display less non-neutralising epitopes than recombinant monomeric gp120, which act to suppress the immune response to target epitopes.

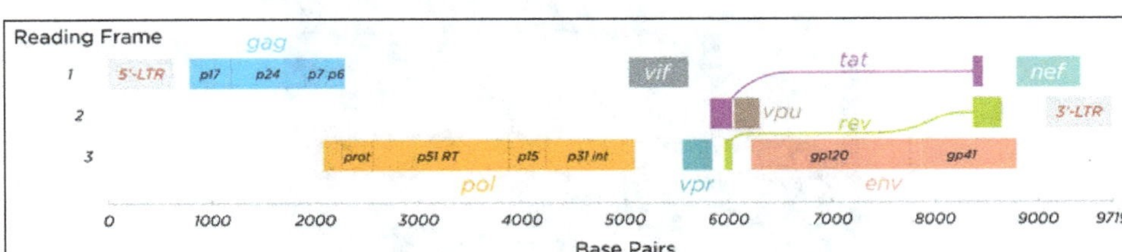

Structure of the RNA genome of HIV-1.

The RNA genome consists of at least seven structural landmarks (LTR, TAR, RRE, PE, SLIP, CRS, and INS), and nine genes (gag, pol, and env, tat, rev, nef, vif, vpr, vpu, and sometimes a tenth tev, which is a fusion of tat, env and rev), encoding 19 proteins. Three of these genes, gag, pol, and env, contain information needed to make the structural proteins for new virus particles. For example, env codes for a protein called gp160 that is cut in two by a cellular protease to form gp120 and gp41. The six remaining genes, tat, rev, nef, vif, vpr, and vpu (or vpx in the case of HIV-2), are regulatory genes for proteins that control the ability of HIV to infect cells, produce new copies of virus (replicate), or cause disease.

The two tat proteins (p16 and p14) are transcriptional transactivators for the LTR promoter acting by binding the TAR RNA element. The TAR may also be processed into microRNAs that regulate the apoptosis genes ERCC1 and IER3. The rev protein (p19) is involved in shuttling RNAs from the nucleus and the cytoplasm by binding to the RRE RNA element. The vif protein (p23) prevents the action of APOBEC3G (a cellular protein that deaminates cytidine to uridine in the single-stranded viral DNA and interferes with reverse transcription). The vpr protein (p14) arrests cell division at G2/M. The nef protein (p27) down-regulates CD4 (the major viral receptor), as well as the MHC class I and class II molecules.

Nef also interacts with SH3 domains. The vpu protein (p16) influences the release of new virus particles from infected cells. The ends of each strand of HIV RNA contain an RNA sequence called a long terminal repeat (LTR). Regions in the LTR act as switches to control production of new viruses and can be triggered by proteins from either HIV or the host cell. The Psi element is involved in viral genome packaging and recognized by gag and rev proteins. The SLIP element (TTTTTT) is involved in the frameshift in the gag-pol reading frame required to make functional pol.

Tropism

The term viral tropism refers to the cell types a virus infects. HIV can infect a variety of immune cells such as CD4⁺ T cells, macrophages, and microglial cells. HIV-1 entry to macrophages and CD4⁺ T cells is mediated through interaction of the virion envelope glycoproteins (gp120) with the CD4 molecule on the target cells' membrane and also with chemokine co-receptors.

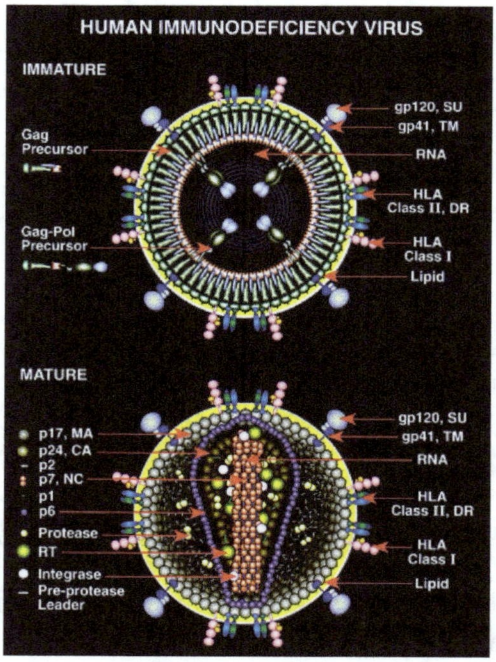

Diagram of the immature and mature forms of HIV.

Macrophage-tropic (M-tropic) strains of HIV-1, or non-syncytia-inducing strains (NSI; now called R5 viruses) use the β-chemokine receptor, CCR5, for entry and are thus able to replicate in both macrophages and CD4⁺ T cells. This CCR5 co-receptor is used by almost all primary HIV-1 isolates regardless of viral genetic subtype. Indeed, macrophages play a key role in several critical aspects of HIV infection. They appear to be the first cells infected by HIV and perhaps the source of HIV production when $CD4^+$ cells become depleted in the patient. Macrophages and microglial cells are the cells infected by HIV in the central nervous system. In the tonsils and adenoids of HIV-infected patients, macrophages fuse into multinucleated giant cells that produce huge amounts of virus.

T-tropic strains of HIV-1, or syncytia-inducing strains (SI; now called X4 viruses) replicate in primary $CD4^+$ T cells as well as in macrophages and use the α-chemokine receptor, CXCR4, for entry.

Dual-tropic HIV-1 strains are thought to be transitional strains of HIV-1 and thus are able to use both CCR5 and CXCR4 as co-receptors for viral entry.

The α-chemokine SDF-1, a ligand for CXCR4, suppresses replication of T-tropic HIV-1 isolates. It does this by down-regulating the expression of CXCR4 on the surface of HIV target cells. M-tropic HIV-1 isolates that use only the CCR5 receptor are termed R5; those that use only CXCR4 are termed X4, and those that use both, X4R5. However, the use of co-receptors alone does not explain viral tropism, as not all R5 viruses are able to use CCR5 on macrophages for a productive infection and HIV can also infect a subtype of myeloid dendritic cells, which probably constitute a reservoir that maintains infection when $CD4^+$ T cell numbers have declined to extremely low levels.

Some people are resistant to certain strains of HIV. For example, people with the CCR5-Δ32 mutation are resistant to infection by the R5 virus, as the mutation leaves HIV unable to bind to this co-receptor, reducing its ability to infect target cells.

Sexual intercourse is the major mode of HIV transmission. Both X4 and R5 HIV are present in the seminal fluid, which enables the virus to be transmitted from a male to his sexual partner. The virions can then infect numerous cellular targets and disseminate into the whole organism. However, a selection processleads to a predominant transmission of the R5 virus through this pathway. In patients infected with subtype B HIV-1, there is often a co-receptor switch in late-stage disease and T-tropic variants that can infect a variety of T cells through CXCR4. These variants then replicate more aggressively with heightened virulence that causes rapid T cell depletion, immune system collapse, and opportunistic infections that mark the advent of AIDS. Thus, during the course of infection, viral adaptation to the use of CXCR4 instead of CCR5 may be a key step in the progression to AIDS. A number of studies with subtype B-infected individuals have determined that between 40 and 50 percent of AIDS patients can harbour viruses of the SI and, it is presumed, the X4 phenotypes.

HIV-2 is much less pathogenic than HIV-1 and is restricted in its worldwide distribution to West Africa. The adoption of "accessory genes" by HIV-2 and its more promiscuous pattern of co-receptor usage (including CD4-independence) may assist the virus in its adaptation to avoid innate restriction factors present in host cells. Adaptation to use normal cellular machinery to enable transmission and productive infection has also aided the establishment of HIV-2 replication in humans. A survival strategy for any infectious agent is not to kill its host, but ultimately become a commensal organism. Having achieved a low pathogenicity, over time, variants that are more successful at transmission will be selected.

Replication Cycle

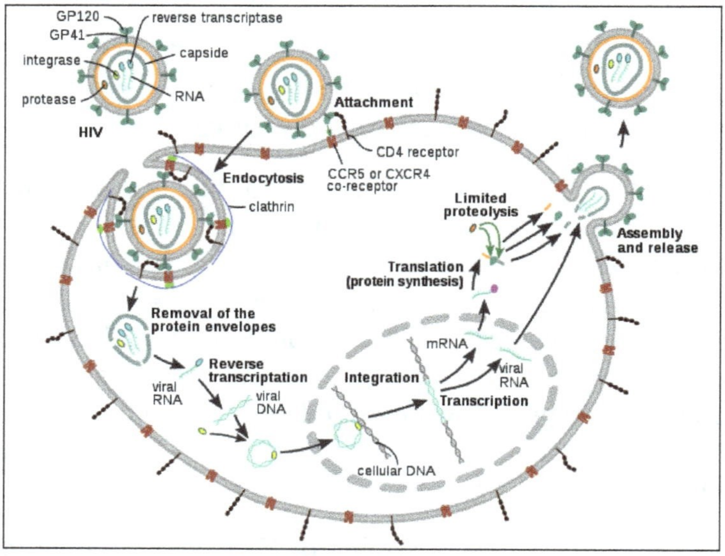

The HIV replication cycle.

Entry to the Cell

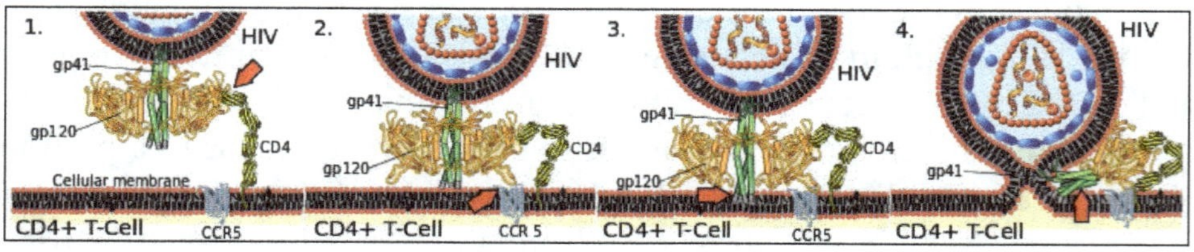

Mechanism of viral entry: 1. Initial interaction between gp120 and CD4. 2. Conformational change in gp120 allows for secondary interaction with CCR5. 3. The distal tips of gp41 are inserted into the cellular membrane. 4. gp41 undergoes significant conformational change; folding in half and forming coiled-coils. This process pulls the viral and cellular membranes together, fusing them.

The HIV virion enters macrophages and CD4$^+$ T cells by the adsorption of glycoproteins on its surface to receptors on the target cell followed by fusion of the viral envelope with the target cell membrane and the release of the HIV capsid into the cell.

Entry to the cell begins through interaction of the trimeric envelope complex (gp160 spike) on the HIV viral envelope and both CD4 and a chemokine co-receptor (generally either CCR5 or CXCR4, but others are known to interact) on the target cell surface. Gp120 binds to integrin $\alpha_4\beta_7$ activating LFA-1, the central integrin involved in the establishment of virological synapses, which facilitate efficient cell-to-cell spreading of HIV-1. The gp160 spike contains binding domains for both CD4 and chemokine receptors.

The first step in fusion involves the high-affinity attachment of the CD4 binding domains of gp120 to CD4. Once gp120 is bound with the CD4 protein, the envelope complex undergoes a structural change, exposing the chemokine receptor binding domains of gp120 and allowing them to interact with the target chemokine receptor. This allows for a more stable two-pronged attachment, which allows the N-terminal fusion peptide gp41 to penetrate the cell membrane. Repeat sequences in gp41, HR1, and HR2 then interact, causing the collapse of the extracellular portion of gp41 into a hairpin shape. This loop structure brings the virus and cell membranes close together, allowing fusion of the membranes and subsequent entry of the viral capsid.

After HIV has bound to the target cell, the HIV RNA and various enzymes, including reverse transcriptase, integrase, ribonuclease, and protease, are injected into the cell. During the microtubule-based transport to the nucleus, the viral single-strand RNA genome is transcribed into double-strand DNA, which is then integrated into a host chromosome.

HIV can infect dendritic cells (DCs) by this CD4-CCR5 route, but another route using mannose-specific C-type lectin receptors such as DC-SIGN can also be used. DCs are one of the first cells encountered by the virus during sexual transmission. They are currently thought to play an important role by transmitting HIV to T cells when the virus is captured in the mucosa by DCs. The presence of FEZ-1, which occurs naturally in neurons, is believed to prevent the infection of cells by HIV.

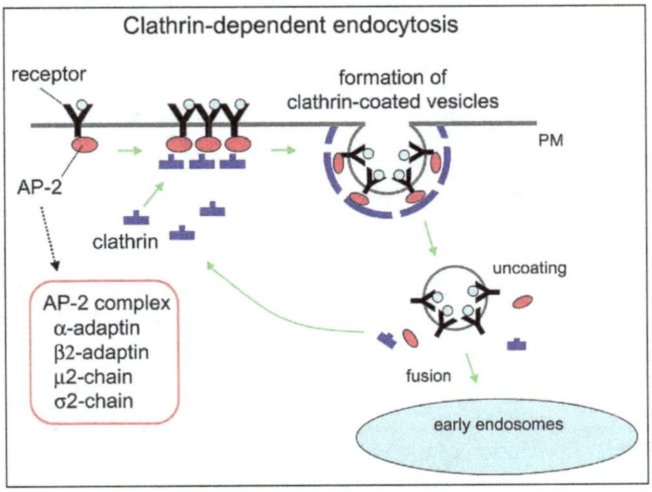

Clathrin-mediated endocytosis.

HIV-1 entry, as well as entry of many other retroviruses, has long been believed to occur exclusively at the plasma membrane. More recently, however, productive infection by pH-independent, clathrin-mediated endocytosis of HIV-1 has also been reported and was recently suggested to constitute the only route of productive entry.

Replication and Transcription

Shortly after the viral capsid enters the cell, an enzyme called reverse transcriptase liberates the positive-sense single-stranded RNA genome from the attached viral proteins and copies it into a complementary DNA (cDNA) molecule. The process of reverse transcription is extremely error-prone, and the resulting mutations may cause drug resistance or allow the virus to evade the body's immune system. The reverse transcriptase also has ribonuclease activity that degrades the viral RNA during the synthesis of cDNA, as well as DNA-dependent DNA polymerase activity that creates a sense DNA from the antisense cDNA. Together, the cDNA and its complement form a double-stranded viral DNA that is then transported into the cell nucleus. The integration of the viral DNA into the host cell's genome is carried out by another viral enzyme called integrase.

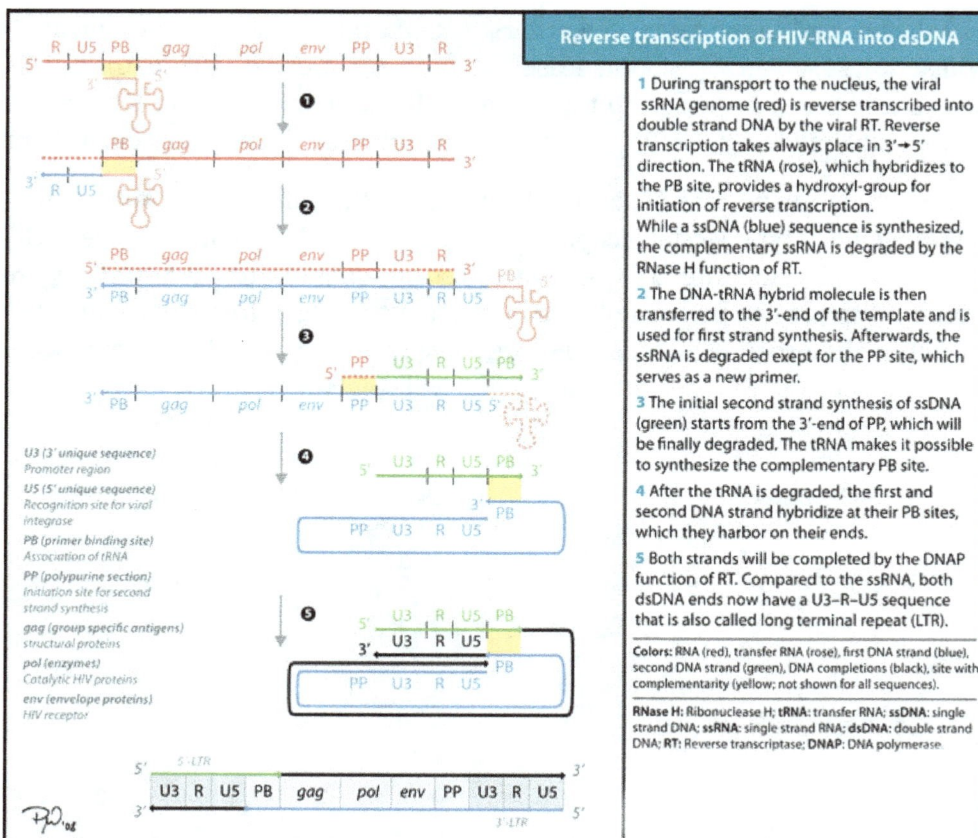

Reverse transcription of the HIV genome into double-stranded DNA.

The integrated viral DNA may then lie dormant, in the latent stage of HIV infection. To actively produce the virus, certain cellular transcription factors need to be present, the most important of which is NF-κB (nuclear factor kappa B), which is upregulated when

T cells become activated. This means that those cells most likely to be targeted, entered and subsequently killed by HIV are those actively fighting infection.

During viral replication, the integrated DNA provirus is transcribed into RNA, some of which then undergo RNA splicing to produce mature messenger RNAs (mRNAs). These mRNAs are exported from the nucleus into the cytoplasm, where they are translated into the regulatory proteins Tat (which encourages new virus production) and Rev. As the newly produced Rev protein is produced it moves to the nucleus, where it binds to full-length, unspliced copies of virus RNAs and allows them to leave the nucleus. Some of these full-length RNAs function as new copies of the virus genome, while others function as mRNAs that are translated to produce the structural proteins Gag and Env. Gag proteins bind to copies of the virus RNA genome to package them into new virus particles.

HIV-1 and HIV-2 appear to package their RNA differently. HIV-1 will bind to any appropriate RNA. HIV-2 will preferentially bind to the mRNA that was used to create the Gag protein itself.

Recombination

Two RNA genomes are encapsidated in each HIV-1 particle. Upon infection and replication catalyzed by reverse transcriptase, recombination between the two genomes can occur. Recombination occurs as the single-strand, positive-sense RNA genomes are reverse transcribed to form DNA. During reverse transcription, the nascent DNA can switch multiple times between the two copies of the viral RNA. This form of recombination is known as copy-choice. Recombination events may occur throughout the genome. Anywhere from two to 20 recombination events per genome may occur at each replication cycle, and these events can rapidly shuffle the genetic information that is transmitted from parental to progeny genomes.

Viral recombination produces genetic variation that likely contributes to the evolution of resistance to anti-retroviral therapy. Recombination may also contribute, in principle, to overcoming the immune defenses of the host. Yet, for the adaptive advantages of genetic variation to be realized, the two viral genomes packaged in individual infecting virus particles need to have arisen from separate progenitor parental viruses of differing genetic constitution. It is unknown how often such mixed packaging occurs under natural conditions.

Bonhoeffer et al. suggested that template switching by reverse transcriptase acts as a repair process to deal with breaks in the single-stranded RNA genome. In addition, Hu and Temin suggested that recombination is an adaptation for repair of damage in the RNA genomes. Strand switching (copy-choice recombination) by reverse transcriptase could generate an undamaged copy of genomic DNA from two damaged single-stranded RNA genome copies. This view of the adaptive benefit of recombination in HIV could explain why each HIV particle contains two complete genomes, rather than one. Furthermore, the view that recombination is a repair process implies that the benefit of repair can occur at each replication cycle, and that this benefit can be realized whether or not the two genomes differ genetically. On the view that recombination in HIV is a repair process, the generation of recombinational variation would be a consequence, but not the cause of, the evolution of template switching.

HIV-1 infection causes chronic inflammation and production of reactive oxygen species. Thus, the HIV genome may be vulnerable to oxidative damages, including breaks in the single-stranded RNA. For HIV, as well as for viruses in general, successful infection depends on overcoming host defensive strategies that often include production of genome-damaging reactive oxygen species. Thus, Michod et al. suggested that recombination by viruses is an adaptation for repair of genome damages, and that recombinational variation is a byproduct that may provide a separate benefit.

Assembly and Release

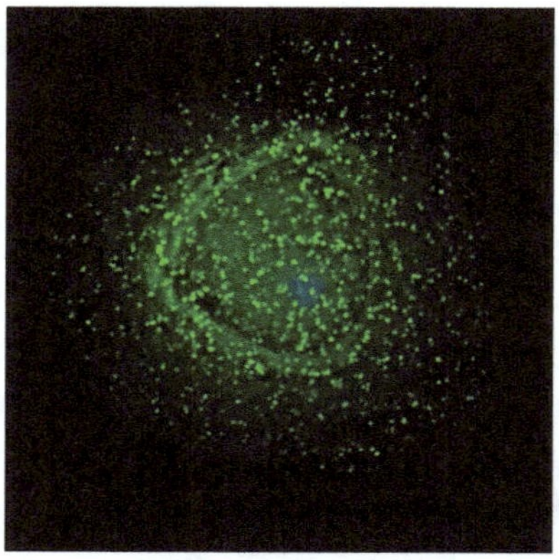

HIV assembling on the surface of an infected macrophage. The HIV virions have been marked with a green fluorescent tag and then viewed under a fluorescent microscope.

The final step of the viral cycle, assembly of new HIV-1 virions, begins at the plasma membrane of the host cell. The Env polyprotein (gp160) goes through the endoplasmic reticulum and is transported to the Golgi apparatus where it is cleaved by furin resulting in the two HIV envelope glycoproteins, gp41 and gp120. These are transported to the plasma membrane of the host cell where gp41 anchors gp120 to the membrane of the infected cell. The Gag (p55) and Gag-Pol (p160) polyproteins also associate with the inner surface of the plasma membrane along with the HIV genomic RNA as the forming virion begins to bud from the host cell. The budded virion is still immature as the gag polyproteins still need to be cleaved into the actual matrix, capsid and nucleocapsid proteins. This cleavage is mediated by the packaged viral protease and can be inhibited by antiretroviral drugs of the protease inhibitor class. The various structural components then assemble to produce a mature HIV virion. Only mature virions are then able to infect another cell.

Spread within the Body

The classical process of infection of a cell by a virion can be called "cell-free spread" to distinguish it from a more recently recognized process called "cell-to-cell spread". In cell-free spread, virus particles bud from an infected T cell, enter the blood or extracellular fluid and then infect another T cell following a chance encounter. HIV can also disseminate by direct transmission from one cell to another by a process of cell-to-cell spread, for which two pathways have been described. Firstly, an infected T cell can transmit virus directly to a target T cell via a virological synapse. Secondly, an antigen-presenting cell (APC), such as a macrophage or dendritic cell, can transmit HIV to T cells by a process that either involves productive infection (in the case of macrophages) or capture and transfer of virions in trans (in the case of dendritic cells). Whichever pathway is used, infection by cell-to-cell transfer is reported to be much more efficient than cell-free virus spread. A number of factors contribute to this increased efficiency, including polarised virus budding towards the site of cell-to-cell contact, close apposition of cells, which minimizes fluid-phase diffusion of virions, and clustering of HIV entry receptors on the target cell towards the contact zone. Cell-to-cell spread is thought to be particularly important in lymphoid tissues where $CD4^+$ T cells are densely packed and likely to interact frequently. Intravital imaging studies have supported the concept of the HIV virological synapse in vivo. The many spreading mechanisms available to HIV contribute to the virus' ongoing replication in spite of anti-retroviral therapies.

Genetic Variability

HIV differs from many viruses in that it has very high genetic variability. This diversity is a result of its fast replication cycle, with the generation of about 10^{10} virions every day, coupled with a high mutation rate of approximately 3×10^{-5} per nucleotide base per cycle of replication and recombinogenic properties of reverse transcriptase.

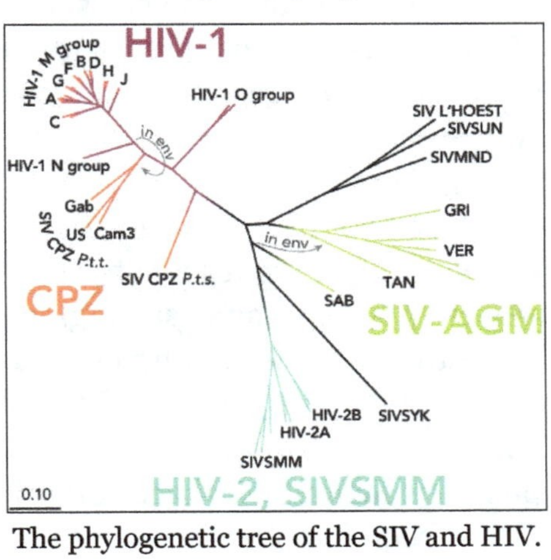

The phylogenetic tree of the SIV and HIV.

This complex scenario leads to the generation of many variants of HIV in a single infected patient in the course of one day. This variability is compounded when a single cell is simultaneously infected by two or more different strains of HIV. When simultaneous infection occurs, the genome of progeny virions may be composed of RNA strands from two different strains. This hybrid virion then infects a new cell where it undergoes replication. As this happens, the reverse transcriptase, by jumping back and forth between the two different RNA templates, will generate a newly synthesized retroviral DNA sequence that is a recombinant between the two parental genomes. This recombination is most obvious when it occurs between subtypes.

The closely related simian immunodeficiency virus (SIV) has evolved into many strains, classified by the natural host species. SIV strains of the African green monkey (SIVagm) and sooty mangabey (SIVsmm) are thought to have a long evolutionary history with their hosts. These hosts have adapted to the presence of the virus, which is present at high levels in the host's blood, but evokes only a mild immune response, does not cause the development of simian AIDS, and does not undergo the extensive mutation and recombination typical of HIV infection in humans.

In contrast, when these strains infect species that have not adapted to SIV ("heterologous" or similar hosts such as rhesus or cynomologus macaques), the animals develop AIDS and the virus generates genetic diversity similar to what is seen in human HIV infection. Chimpanzee SIV (SIVcpz), the closest genetic relative of HIV-1, is associated with increased mortality and AIDS-like symptoms in its natural host. SIVcpz appears to have been transmitted relatively recently to chimpanzee and human populations, so their hosts have not yet adapted to the virus. This virus has also lost a function of the nef gene that is present in most SIVs. For non-pathogenic SIV variants, nef suppresses T cell activation through the CD3 marker. Nef's function in non-pathogenic forms of SIV is to downregulate expression of inflammatory cytokines, MHC-1, and signals that affect T cell trafficking. In HIV-1 and SIVcpz, nef does not inhibit T-cell activation and it has lost this function. Without this function, T cell depletion is more likely, leading to immunodeficiency.

Three groups of HIV-1 have been identified on the basis of differences in the envelope (env) region: M, N, and O. Group M is the most prevalent and is subdivided into eight subtypes (or clades), based on the whole genome, which are geographically distinct. The most prevalent are subtypes B (found mainly in North America and Europe), A and D (found mainly in Africa), and C (found mainly in Africa and Asia); these subtypes form branches in the phylogenetic tree representing the lineage of the M group of HIV-1. Co-infection with distinct subtypes gives rise to circulating recombinant forms (CRFs). In 2000, the last year in which an analysis of global subtype prevalence was made, 47.2% of infections worldwide were of subtype C, 26.7% were of subtype A/CRF02_AG, 12.3% were of subtype B, 5.3% were of subtype D, 3.2% were of CRF_AE, and the remaining 5.3% were composed of other subtypes and CRFs. Most HIV-1 research is focused on subtype B; few laboratories focus on the other subtypes. The existence of a fourth group, "P", has been hypothesised based on a virus isolated in 2009.

The strain is apparently derived from gorilla SIV (SIVgor), first isolated from western lowland gorillas in 2006.

HIV-2's closest relative is SIVsm, a strain of SIV found in sooty mangabees. Since HIV-1 is derived from SIVcpz, and HIV-2 from SIVsm, the genetic sequence of HIV-2 is only partially homologous to HIV-1 and more closely resembles that of SIVsm.

Diagnosis

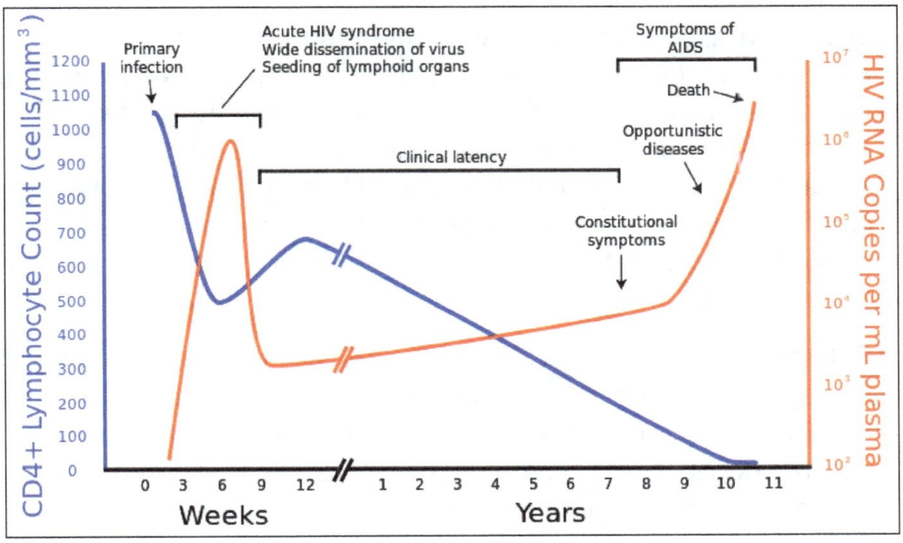

A generalized graph of the relationship between HIV copies (viral load) and CD4 counts over the average course of untreated HIV infection; any particular individual's disease course may vary considerably.

- $CD4^+$ T cell count (cells per µL).
- HIV RNA copies per mL of plasma.

Many HIV-positive people are unaware that they are infected with the virus. For example, in 2001 less than 1% of the sexually active urban population in Africa had been tested, and this proportion is even lower in rural populations. Furthermore, in 2001 only 0.5% of pregnant women attending urban health facilities were counselled, tested or receive their test results. Again, this proportion is even lower in rural health facilities. Since donors may therefore be unaware of their infection, donor blood and blood products used in medicine and medical research are routinely screened for HIV.

HIV-1 testing is initially done using an enzyme-linked immunosorbent assay (ELISA) to detect antibodies to HIV-1. Specimens with a non-reactive result from the initial ELISA are considered HIV-negative, unless new exposure to an infected partner or partner of unknown HIV status has occurred. Specimens with a reactive ELISA result are retested in duplicate. If the result of either duplicate test is reactive, the specimen is reported as repeatedly reactive and undergoes confirmatory testing with a more specific supplemental test (e.g., a polymerase chain reaction (PCR), western blot or, less commonly, an immunofluorescence assay (IFA)). Only specimens that are repeatedly reactive by ELISA and positive by IFA or PCR or reactive by western blot are considered HIV-positive and indicative of HIV infection. Specimens that are repeatedly ELISA-reactive occasionally provide an indeterminate western blot result, which may be either an incomplete antibody response to HIV in an infected person or nonspecific reactions in an uninfected person.

Although IFA can be used to confirm infection in these ambiguous cases, this assay is not widely used. In general, a second specimen should be collected more than a month later and retested for persons with indeterminate western blot results. Although much less commonly available, nucleic acid testing (e.g., viral RNA or proviral DNA amplification method) can also help diagnosis in certain situations. In addition, a few tested specimens might provide inconclusive results because of a low quantity specimen. In these situations, a second specimen is collected and tested for HIV infection.

Modern HIV testing is extremely accurate, when the window period is taken into consideration. A single screening test is correct more than 99% of the time. The chance of a false-positive result in a standard two-step testing protocol is estimated to be about 1 in 250,000 in a low risk population. Testing post-exposure is recommended immediately and then at six weeks, three months, and six months.

The latest recommendations of the US Centers for Disease Control and Prevention (CDC) show that HIV testing must start with an immunoassay combination test for HIV-1 and HIV-2 antibodies and p24 antigen. A negative result rules out HIV exposure, while a positive one must be followed by an HIV-1/2 antibody differentiation immunoassay to detect which antibodies are present. This gives rise to four possible scenarios:

- HIV-1 (+) and HIV-2 (−): HIV-1 antibodies detected.
- HIV-1 (−) and HIV-2 (+): HIV-2 antibodies detected.
- HIV-1 (+) and HIV-2 (+): Both HIV-1 and HIV-2 antibodies detected.
- HIV-1 (−) or indeterminate and HIV-2 (−): Nucleic acid test must be carried out to detect the acute infection of HIV-1 or its absence.

Treatment and Transmission

The management of HIV/AIDS normally includes the use of multiple antiretroviral drugs. In many parts of the world, HIV has become a chronic condition in which progression to AIDS is increasingly rare.

HIV latency, and the consequent viral reservoir in $CD4^+$ T cells, dendritic cells, as well as macrophages, is the main barrier to eradication of the virus.

It is important to note that although HIV is highly virulent, transmission does not occur through sex when an HIV-positive person has a consistently undetectable viral load (<50 copies/ml) due to anti-retroviral treatment. Previously it was said the chance of transmission was "very low" or "negligible" (The "Swiss Statement"). However, following multiple studies, it is now clear that the chance of passing on HIV through sex is zero where the HIV-positive person has a consistently undetectable viral load; this is known as U=U, "Undetectable=Untransmittable", also phrased as "can't pass it on". The studies demonstrating U=U are: Opposites Attract, PARTNER 1, PARTNER 2, (for male-male couples) and HPTN052 (for heterosexual couples) when "the partner living with HIV had a durably suppressed viral load." In these studies, couples where one partner was HIV positive and one partner was HIV negative were enrolled and regular HIV testing completed. In total from the four studies, 4097 couples were enrolled over four continents and 151,880 acts of condomless sex were reported; there were zero phylogenetically linked transmissions of HIV where the positive partner had an undetectable viral load. Following this, the U=U consensus statement advocating the use of "zero risk" was signed by hundreds of individuals and organisations, including the US CDC, British HIV Association and The Lancet medical journal. The importance of the final results of the PARTNER 2 study were described by the medical director of the Terrence Higgins Trust as "impossible to overstate," while lead author Alison Rodger declared that the message that "undetectable viral load makes HIV untransmittable can help end the HIV pandemic by preventing HIV transmission. The authors summarised their findings in The Lancet as follows:

> "Our results provide a similar level of evidence on viral suppression and HIV transmission risk for gay men to that previously generated for heterosexual couples and suggest that the risk of HIV transmission in gay couples through condomless sex when HIV viral load is suppressed is effectively zero. Our findings support the message of the U=U (undetectable equals untransmittable) campaign, and the benefits of early testing and treatment for HIV."

This result is consistent with the conclusion presented by Anthony S. Fauci, the Director of the National Institute of Allergy and Infectious Diseases for the U.S. National Institutes of Health, and his team in a viewpoint in the Journal of the American Medical Association, that U=U is an effective HIV prevention method when an undetectable viral load is maintained.

Human Parainfluenza Viruses

Human parainfluenza viruses (HPIVs) are the viruses that cause human parainfluenza. HPIVs are a paraphyletic group of four distinct single-stranded RNA viruses belonging to the Paramyxoviridae family. These viruses are closely associated with both human and veterinary disease. Virions are approximately 150–250 nm in size and contain negative sense RNA with a genome encompassing about 15,000 nucleotides.

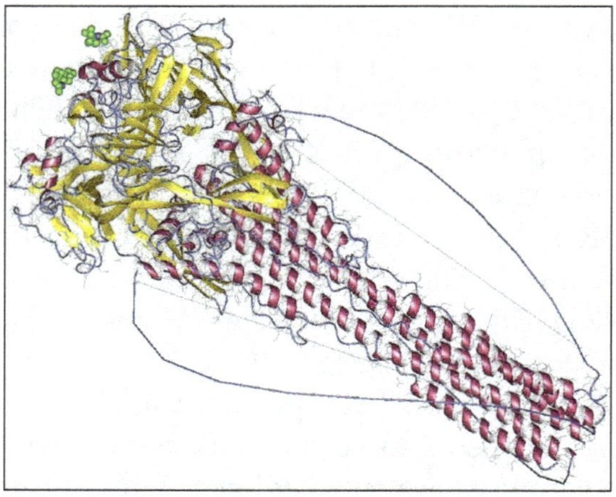

Fusion glycoprotein trimer, Human parainfluenza virus 3 (hPIV3).

The viruses can be detected via cell culture, immunofluorescent microscopy, and PCR. HPIVs remain the second main cause of hospitalisation in children under 5 years of age suffering from a respiratory illness (only Human orthopneumovirus causes more respiratory hospitalisations for this age group).

Classification

The first HPIV was discovered in the late 1950s. The taxonomic division is broadly based on antigenic and genetic characteristics, forming four major serotypes or clades, which today are considered distinct viruses. These include:

Virus	GenBank acronym	NCBI taxonomy	
Human parainfluenza virus type 1	HPIV-1	12730	Most common cause of croup
Human parainfluenza virus type 2	HPIV-2	11212	Causes croup and other upper and lower respiratory tract illnesses
Human parainfluenza virus type 3	HPIV-3	11216	Associated with bronchiolitis and pneumonia
Human parainfluenza virus type 4	HPIV-4	11203	Includes subtypes 4a and 4b

HPIVs belong to two genera: Respirovirus (HPIV-1 and HPIV-3) and Rubulavirus (HPIV-2 and HPIV-4).

Viral Structure and Organisation

hPIVs are characterised by producing enveloped virions and containing single stranded negative sense RNA. Non-infectious virions have also been reported to contain RNA with positive polarity. HPIV genomes are about 15,000 nucleotides in length and encode six key structural proteins.

The structural gene sequence of HPIVs is as follows: 3′-NP-P-M-F-HN-L-5′.

Structural protein	Location	Function
Hemagglutinin-neuraminidase (HN)	Envelope	Attachment and cell entry
Fusion Protein (F)	Envelope	Fusion and cell entry
Matrix Protein (M)	Within the envelope	Assembly
Nucleoprotein (NP)	Nucleocapsid	Forms a complex with the RNA genome
Phosphoprotein (P)	Nucleocapsid	Forms as part of RNA polymerase complex
Large Protein (L)	Nucleocapsid	Forms as part of RNA polymerase complex

With the advent of reverse genetics, it has been found that the most efficient human parainfluenza viruses (in terms of replication and transcription) have a genome nucleotide total that is divisible by the number 6. This has led to the "rule of six" being coined. Exceptions to this rule have been found and its exact advantages are not fully understood.

Electrophoresis has shown that the molecular weight (MW) of the proteins for the four HPIVs are similar (with the exception of the phosphoprotein, which shows significant variation).

Viral Entry and Replication

Viral replication is initiated only after successful entry into a cell by attachment and fusion between the virus and the host cell lipid membrane. Viral RNA (vRNA) is initially associated with nucleoprotein (NP), phosphoprotein (P) and the large protein (L). The hemagglutinin–neuraminidase (HN) is involved with viral attachment and thus hemadsorption and hemagglutination. Furthermore, the fusion (F) protein is important in aiding the fusion of the host and viral cellular membranes, eventually forming syncytia.

Initially the F protein is in an inactive form (F_0) but can be cleaved by proteolysis to form its active form, F_1 and F_2, linked by di-sulphide bonds. Once complete, this is followed by the HPIV nucleocapsid entering the cytoplasm of the cell. Subsequently, genomic transcription occurs using the viruses own 'viral RNA-dependent RNA polymerase' (L protein). The cell's own ribosomes are then tasked with translation, forming the viral proteins from the viral mRNA.

Towards the end of the process, (after the formation of the viral proteins) the replication of the viral genome occurs. Initially, this occurs with the formation of a positive-sense RNA (intermediate step, necessary for producing progeny), and finally, negative-sense RNA is formed which is then associated with the nucleoprotein. This may then be either packaged and released from the cell by budding or used for subsequent rounds of transcription and replication.

The observable and morphological changes that can be seen in infected cells include the enlargement of the cytoplasm, decreased mitotic activity and 'focal rounding', with the potential formation of multi-nucleate cells (syncytia).

The pathogenicity of HPIVs is mutually dependent on the viruses having the correct accessory proteins that are able to elicit anti-interferon properties. This is a major factor in the clinical significance of disease.

Host Range

The main host remains the human. However, infections have been induced in other animals (both under natural and experimental situations), although these were always asymptomatic.

Clinical Significance

It is estimated that there are 5 million children with lower respiratory infections (LRI) each year in the United States alone. HPIV-1, HPIV-2 and HPIV-3 have been linked with up to a third of these infections. Upper respiratory infections (URI) are also important in the context of HPIV, however, they are caused to a lesser extent by the virus. The highest rates of serious HPIV illnesses occur among young children, and surveys have shown that about 75% of children aged 5 or older have antibodies to HPIV-1.

For infants and young children, it has been estimated that about 25% will develop "clinically significant disease".

Repeated infection throughout the life of the host is not uncommon and symptoms of later breakouts include upper respiratory tract illness, such as cold and a sore throat. The incubation period for all four serotypes is 1 to 7 days. In immunosuppressed people, parainfluenza virus infections can cause severe pneumonia, which can be fatal.

HPIV-1 and HPIV-2 have been demonstrated to be the principal causative agent behind croup (laryngotracheobronchitis), which is a viral disease of the upper airway and is mainly problematic in children aged 6–48 months of age. Biennial epidemics starting in Autumn are associated with both HPIV-1 and 2; however, HPIV-2 can also have yearly outbreaks. Additionally, HPIV-1 tends to cause biennial outbreaks of croup in the Fall. In the United States, large peaks have presently been occurring during odd-numbered years.

HPIV-3 has been closely associated with bronchiolitis and pneumonia and principally targets those aged <1 year.

HPIV-4 remains infrequently detected. However, it is now believed to be more common than previously thought, but is less likely to cause severe disease. By the age of 10, the majority of children are sero-positive for HPIV-4 infection which may be indicative of a large proportion of asymptomatic or mild infections.

Important epidemiological factors that are associated with a higher risk of infection and mortality are those who are immuno-compromised and may be taken ill with more extreme forms of LRI. Associations between HPIVs and neurologic disease are known; for example, hospitalisation with certain HPIVs has a strong association with febrile seizures. HPIV-4B has the strongest association (up to 62%) followed by hPIV-3 and -1.

HPIVs have also been linked with rare cases of virally caused meningitis and Guillain–Barré syndrome.

HPIVs are spread from person to person ('horizontal transmission') by contact with infected secretions through respiratory droplets or contaminated surfaces or objects. Infection can occur when infectious material contacts mucous membranes of the eyes, mouth, or nose, and possibly through the inhalation of droplets generated by a sneeze or cough. HPIVs can remain infectious in airborne droplets for over an hour.

Overall, HPIVs remain best known for its effects on the respiratory system and this appears to be where the majority of the focus has been upon.

Airway Inflammation

The inflammation of the airway is a common attribute of HPIV infection. It is believed to occur due to the large scale up-regulation of inflammatory cytokines. Common cytokines, expected to be up-regulated, include IFN–α, various other interleukins (IL–2, IL-6) and TNF–α. Various chemokines and inflammatory proteins are also believed to be associated with the common symptoms of HPIV infection.

Recent evidence seems to suggest that virus-specific antibodies (IgE) may be responsible for mediating large-scale releases of histamine in the trachea, which are believed to cause croup.

Immunology

The body's primary defense against HPIV infection remains humoral immunity. This is mainly directed against surface proteins which can be found on the virus. In particular the proteins HN and F prove to be most immunogenic in terms of stimulating the immune system.

Recently the importance of the 'cell mediated immune system' has been scrutinized. Reports have shown that those with defective adaptive immune responses are at a higher risk of severe infection.

Morbidity and Mortality

Mortality caused by HPIVs in developed regions of the world remains rare. Where mortality has occurred, it is principally in the three core risk groups (very young, elderly and immuno-compromised). Long term changes can however be associated with airway remodelling and are believed to be a significant cause of morbidity. The exact associations between HPIVs and diseases such as chronic obstructive pulmonary disease (COPD) are still being investigated.

In developing regions of the world, preschool children remain the highest mortality risk group. Mortality may be a consequence of primary viral infection or secondary problems such as bacterial infection. Predispositions, such as malnutrition and other deficiencies may further elevate the chances of mortality associated with infection.

Overall, LRI's cause approximately 25–30% of total deaths in preschool children in the developing world. HPIVs is believed to be associated with 10% of all LRI cases, thus remaining a significant cause of mortality.

Risk Factors

Numerous factors have been suggested and linked to a higher risk of acquiring the infection, inclusive of malnutrition, vitamin A deficiency, absence of breastfeeding during the early stages of life, environmental pollution and overcrowding.

Prevention

Despite decades of research, no vaccines currently exist.

Recombinant technology has however been used to target the formation of vaccines for HPIV-1, -2 and -3 and has taken the form of several live-attenuated intranasal vaccines. Two vaccines in particular were found to be immunogenic and well tolerated against HPIV-3 in phase I trials. HPIV-1 and -2 vaccine candidates remain less advanced.

Vaccine techniques which have been used against HPIVs are not limited to intranasal forms, but also viruses attenuated by cold passage, host range attenuation, chimeric construct vaccines and also introducing mutations with the help of reverse genetics to achieve attenuation.

Maternal antibodies may offer some degree of protection against HPIVs during the early stages of life via the colostrum in breast milk.

Medication

Ribavirin is one medication which has shown good potential for the treatment of HPIV-3 given recent in-vitro tests (in-vivo tests show mixed results). Ribavirin is a broad-spectrum antiviral, and is currently being administered to those who are severely immuno-compromised, despite the lack of conclusive evidence for its benefit. Protein inhibitors and novel forms of medication have also been proposed to relieve the symptoms of infection.

Furthermore, antibiotics may be used if a secondary bacterial infection develops. Corticosteroid treatment and nebulizers are also a first line choice against croup if breathing difficulties ensue.

Interactions with the Environment

Parainfluenza viruses last only a few hours in the environment and are inactivated by soap and water. Furthermore, the virus can also be easily destroyed using common hygiene techniques and detergents, disinfectants and antiseptics.

Environmental factors which are important for HPIV survival are pH, humidity, temperature and the medium within which the virus is found. The optimal pH is around the physiologic pH values (7.4 to 8.0), whilst at high temperatures (above 37 °C) and low humidity, infectivity reduces.

The majority of transmission has been linked to close contact, especially in nosocomial infections. Chronic care facilities and doctors' surgeries are also known to be transmission 'hotspots' with transmission occurring via aerosols, large droplets and also fomites (contaminated surfaces). The exact infectious dose remains unknown.

Economic Burden

In the poorest regions of the world, HPIV infection can be measured in terms of mortality. In the developed world where mortality remains rare, the economic costs of the infection can be estimated. Estimates from the US are suggestive of a cost (based on extrapolation) in the region of $200 million per annum.

Hepatitis B Virus

Hepatitis B virus is a hepadnavirus. The discovery by Baruch Blumberg and colleagues of the Australia antigen, which would later be identified as the hepatitis B virus (HBV) surface antigen, was a major breakthrough towards improving global health. For decades prior to Blumberg's discovery, an unknown virus in blood and plasma samples had been the suspected cause of post-transfusion hepatitis. Recognition that the Australia antigen was a marker of viral hepatitis facilitated the generation of a blood-screening protocol that led to a two- to three-fold reduction in the incidence of post-transfusion hepatitis, with the remaining cases likely caused by hepatitis C virus (HCV), which would not be identified for another 23 years.

Retrospective studies also associated the presence of the Australia antigen with chronic liver diseases such as cirrhosis and hepatocellular carcinoma (HCC). Finally, the discovery of the Australia antigen also facilitated the eventual development of a vaccine that has greatly reduced the global burden of HBV infection, and Baruch Blumberg was awarded the 1976 Nobel Prize in Medicine for his discovery of HBV.

Despite the decades of work between the discovery of HBV and our current understanding of the virus, many aspects of the HBV life cycle and pathogenesis remain unclear. The fact that it is estimated that as much as a third of the world's population has been infected with HBV at some point, that roughly 5% of the population (350-500 million people) are chronically infected with the virus, and that about 800,000 people die annually from acute or chronic HBV-related consequences underscores the importance of a more complete understanding of HBV biology and pathogenesis.

Hepadnaviridae

While the discovery of human HBV, hereafter denoted as HBV, occurred in the 1960s, recent research has shown that hepatitis B viruses have actually been present since the time of the dinosaurs. In fact, the earliest known hepatitis B virus is approximately 82 million years old and was identified from the DNA of infected birds from the Mesozoic period. Although multiple theories of the origins of HBV exist, it appears that the infection of mammals is a much more recent event. The jump into humans, in particular, may have been only about 40,000 years ago. Despite the evolutionary timeline, modern HBV is remarkably similar to these ancient hepatitis B viruses.

The present day Hepadnaviridae family is a group of small, hepatotropic, DNA viruses that are divided into two distinct genera based on their divergent genomic sequences and narrow host range of infection. The avihepadnaviruses, such as duck HBV (DHBV) and heron HBV, infect birds. In contrast, the orthohepadnaviruses infect mammals and include HBV and woodchuck hepatitis virus (WHV), among others. Each member of the Hepadnaviridae family is primarily species specific. For example, the only non-human hosts of HBV are chimpanzee and treeshrew, each of which can be experimentally infected. Additionally, a primate virus similar to HBV, called woolly monkey HBV, has been identified in woolly monkeys and designated as the prototype of a new species of hepatitis B-like viruses. A maximum of 40% sequence divergence exists between orthohepadnaviruses, while only 20% sequence divergence exists among avihepadnaviruses; however, little to no homology exists between the two genera. All mammalian HBV encode an X protein, which has been shown to be required for viral replication and has oncogenic properties. This X protein is either lacking or highly divergent in avian viruses, and the acquisition of this X protein could have been an essential factor for the evolution of hepadnaviruses from avian into mammalian hosts.

While significant genomic diversity exists between viral species and particularly between the hepadnaviridae genera, all hepadnaviruses share a large number of common features. Among these, all members have an extremely small (3.0-3.3 kb) and compact DNA genome that encodes overlapping open reading frames (ORFs). Additionally, all hepadnaviruses use a genome replication strategy in which the virus replicates its DNA genome by reverse transcription of an RNA intermediate using the reverse-transcriptase activity of the viral polymerase. Hepadnaviruses are also distinguished from nearly all other viruses utilizing reverse transcription for viral replication by a number of unique features, including envelopment of a DNA genome, rather than RNA, and the fact that integration of the hepadnavirus DNA genome into the host-cell genome is not required for viral replication. These features, common to all members of the hepadnavirus family, contributed to the designation of Hepadnaviridae as a distinct family of viruses.

HBV

Studies have identified a minimum of eight HBV genotypes, designated A-H, with genetic differences greater than 8%, but less than 17% between each genotype. Two additional potential genotypes have been described. Genotype I has genetic divergence around 8% with a strong homology to genotype C, making its classification as a distinct genotype more controversial than that of the more well-accepted genotypes. A potential 10th genotype, genotype J, has also been described recently and is likely the result of recombination of genotype C and gibbon HBV.

There is a distinct distribution of HBV genotypes within specific populations and geographic locations. Similarly, there is an association between genotype and disease severity and outcome. In the United States, where chronic HBV infection is relatively uncommon, each genotype is present, though not at equal levels. Within the United States population, genotypes A and D are most prevalent overall, and the distribution of genotypes can be further divided based on ethnicity. For example, genotype C is most common in Asian Americans, which correlates with the prevalence of this genotype in much of Asia. This is significant because genotype C has been associated with a more severe disease and a lower response rate to interferon therapy.

HBV Genome Organization

HBV has a small (3.2 kb), partially double-stranded, relaxed-circular DNA genome that encodes four overlapping ORFs. The largest ORF encodes the viral polymerase, which also has reverse transcriptase (RT) activity that generates the first strand of the DNA genome from an RNA intermediate. The second largest ORF encodes the three viral envelope proteins: large (L-), middle (M-), and small (S-) surface antigen (HBsAg). Another ORF encodes precore, also referred to as HBV E antigen (HBeAg), and the core protein, which makes up the viral capsid. Finally, the smallest ORF encodes the HBV X protein (HBx), a small regulatory protein that has been shown to be required for HBV replication both in vitro and in vivo. The viral ORFs are encoded in distinct capped and polyadenylated RNAs that can be divided into genomic and subgenomic transcripts. The subgenomic transcripts act only as templates for HBV proteins and consist of the

0.7 kb transcript, which encodes HBx, and the 2.1 kb and 2.4 kb HBsAg transcripts encoding M- and S-HBsAg, and L-HBsAg, respectively. The genomic transcripts act as mRNAs for precore, core, and polymerase. The genomic transcript that encodes both core and polymerase is multifunctional and referred to as pregenomic RNA (pgRNA). The pgRNA is the template for HBV replication and is reverse transcribed to generate the HBV DNA genome. As the viral genome is only 3.2 kb and the pgRNA is 3.5 kb, the pgRNA is a greater than unit length, terminally redundant copy of the viral genome. All HBV RNA transcripts share the same polyadenylation site, and each of the smaller transcripts makes up the 3' end of each of the larger transcripts. This means that the sequence of the HBx transcript is contained at the 3' end of all HBV mRNA transcripts, while the largest transcript is the only viral transcript to contain sequence that is not shared with the other transcripts.

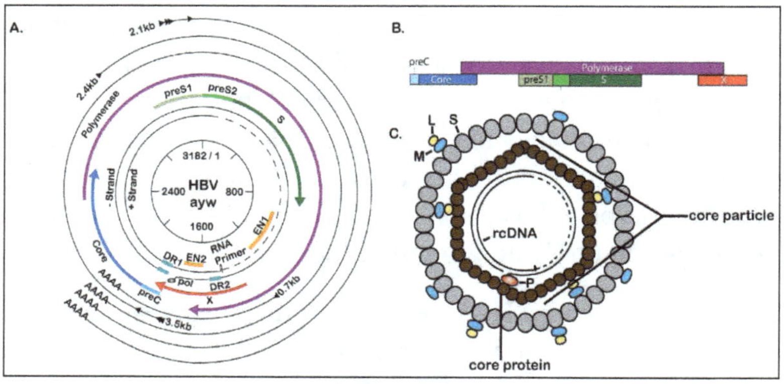

Molecular biology of hepatitis B virus (HBV). (A) Scaled depiction of the HBV (genotype ayw) genome. Internal circle shows genomic position relative to EcoRI site at position 1. Partially double-stranded genome is depicted with attached RNA primer and polymerase protein. Open reading frames (ORFs) are indicated by the thicker, colored lines. The outermost black circles represent the viral transcripts with the shared polyadenylation site; (B) schematic representation of the overlapping nature of the HBV ORFs; (C) the mature HBV virion (Dane particle) consists of two main parts: a nucleocapsid (or core particle) consisting of a partially double-stranded DNA genome bound to polymerase (P) and encapsidated by dimers of core protein, and a viral envelope consisting primarily of S-HBsAg (S), with an intermediate amount of M-HBsAg (M) and lower levels of L-HBsAg (L).

Transcription of HBV RNAs is driven from specific promoter sequences within the viral genome. At least some of the hepatotropic restriction of HBV can be attributed to transcriptional activation by hepatocyte-specific transcription factors. For example, activation of the Enhancer I/HBx promoter is a required first step in viral transcription, as this is believed to enhance transcription from downstream promoters. A number of the transcription factors that have been mapped to the EN1/HBx promoter are liver specific, including hepatocyte nuclear factor (HNF) 1, HNF3, and HNF4. Many of the transcription factor binding sites that have been identified within the 4 promoter regions of

HBV are for transcription factors that are activated by HBV proteins, oftentimes HBx, implying a specific cascade of transcription.

HBV Proteins

The HBV genome encodes seven proteins: HBx, core, polymerase, L-, M-, and S-HBsAg, and precore/HBeAg. Of these proteins, HBx is a non-structural regulatory protein, HBeAg is not incorporated into virions and is independently secreted from the cells, the polymerase is responsible for genome replication, and the core and HBsAg proteins form the structural aspects of the virion.

E Antigen

HBeAg is the final product of post-translational processing of the translated pre-core ORF. As one of the proteins encoded by a genomic transcript, the genomic promoter drives its expression. The HBeAg ORF encodes an endoplasmic reticulum (ER) targeting sequence that co-translationally traffics the peptide to the ER, where the protein is processed to the final 15 kD HBeAg that is secreted from HBV-infected cells.

The function of HBeAg remains incompletely defined. Multiple groups have hypothesized that HBeAg can facilitate HBV immune evasion, and studies with HBc/HBeAg-transgenic (tg) mice crossed with T cell receptor (TCR)-Tg mice expressing receptors for the HBc/HBeAgs specifically suggest that a function of HBeAg is to suppress the immune response to the HBV core protein. The secretion of a viral marker that is not present in the HBV infectious virion may help to dampen the neutralizing immune response by diverting this response away from infectious viral particles. From a diagnostic perspective, HBeAg is an important marker of HBV replication, and the levels of serum HBeAg are generally considered to correlate with viral titer. In fact, HBeAg seroconversion is considered an important aspect of the transition to the inactive carrier state of infection.

Surface Antigens

HBV encodes three envelope proteins, or surface antigens, that make up the viral envelope: large (L), middle (M), and small (S) surface antigen. The smallest envelope protein, or S (24 kD), is 226 amino acids (aa) in length and makes up a shared C-terminal region of the two longer envelope proteins. The M protein (31 kD) contains the S sequence with a 55aa N-terminal extension known as preS2. Expression of the M- and S-encoding mRNA is driven by the S promoter, with translation initiating from an upstream (M) or downstream (S) AUG. The L protein (39 kD), the largest of the envelope proteins, contains S, preS2, and an additional 108aa or 119aa (depending on the genotype) N-terminal extension known as preS1. L-HBsAg is encoded by its own mRNA transcript that is controlled by the preS1 promoter.

The envelope proteins are synthesized at the ER, where they attain their transmembrane configuration. Because all three proteins contain an identical C-terminal sequence, the transmembrane topology of this region is the same across all three proteins. Specifically, an N-terminal signal sequence initiates insertion of S into the ER membrane, followed by another signal that pushes the downstream peptide sequence into the ER lumen. The sequence upstream of this signal remains in the cytosol, with the signal domain itself acting as a transmembrane anchor domain. This orientation forms two loops; one loop, between aa 23-79, remains on the cytosolic side, while the other loop, between aa 99-169, remains in the ER lumen. Importantly, the luminal loop contains the major conformational epitope of HBsAg and is glycosylated in nearly half of all S-protein moieties. Once budding of the membrane occurs, these epitopes are on the outer surface of the viral particles. The topology of M is identical to that of S, except for the presence of preS2 within the ER lumen.

A major characteristic of L is that it exists in two conformations that vary in the localization of the N-terminal domain. In the first conformation of L, the preS1 and preS2 domains are present in the cytosol. This conformation of L is essential for binding of capsids and for the assembly of HBV virions. In the second conformation of L, the N-terminus is located in the ER lumen and, as a result, exposed on the surface of viral particles. Thus, this conformation of L plays a role in the infection of hepatocytes. The conformational change is facilitated by interactions of molecular chaperones Hsc70/Hsp40 and BiP with L; however, the exact details of the mechanism underlying this step are not yet understood. The preS1 domain contains the receptor-binding region for HBV, thus it needs to be exposed out of the cell. A myristylated peptide containing a portion of the N-terminal preS1 region is sufficient to inhibit infection and is currently being developed as a therapeutic.

The main function of the surface antigen proteins is to form the HBV envelope. Three different forms of viral particles are secreted from an HBV-infected cell as a result of the unequal expression of each of the three surface antigens. S-HBsAg is the highest expressed of the three envelope proteins and makes up the majority of the viral envelope. Intact, infectious HBV virions, called Dane particles, also include M-HBsAg and L-HBsAg. In addition, an HBV-infected cell produces non-infectious subviral particles (SVP) made primarily of S-HBsAg containing varying (but much lower) amounts of M-HBsAg and little to no L-HBsAg. These SVPs can reach a concentration 10,000-fold higher than infectious HBV particles in the serum of an infected individual. SVPs are produced in two forms: 25 nm spheres, which are almost exclusively made up of S-HBsAg, and 22 nm filaments, which are made up primarily of S-HBsAg, with some M-HBsAg and potentially small amounts of L-HBsAg. It is currently unknown why HBV produces SVPs in such excess compared to the level of infectious virions, but multiple hypotheses have been proposed. For example, it has been suggested that the excess SVPs act to divert neutralizing antibodies away from infectious particles and that SVPs play a role in inducing the immune tolerance required to sustain a long-term chronic infection. A study of DHBV SVPs showed that the SVP-to-infectious-particle ratio plays a role in determining the efficiency of hepatocyte infection, with SVPs acting to either enhance or inhibit infection based on the ratio of SVP-to-infectious-particles.

Core Protein

The 21 kD HBV core protein, or HBcAg, is the organizing framework for the virion. When expressed in cells, core mainly exists as soluble dimers, or in T = 3 or T = 4 icosahedral capsids. About 95% of mature nucleocapsids isolated from Dane particles contain T = 4 capsids made up of 120 core dimers, with the remaining 5% being the smaller T = 3 with 90 dimers. Core is translated from the pgRNA and the first 149aa of core form the assembly domain, which is sufficient for in vitro formation of capsids that are indistinguishable from capsids isolated from Dane particles. The remaining 34-36aa makes up the arginine-rich C-terminal domain (CTD); phosphorylation of various aa in the CTD regulates multiple stages of the HBV life cycle.

While the best-described role for core protein is assembling the nucleocapsid, the results of recent studies also suggest that the core protein does more than simply act as an inert container for the HBV genome. In fact, core protein binds to HBV covalently closed circular DNA (cccDNA), potentially to regulate spacing of nucleosomes on cccDNA; cccDNA is a nuclear-localized replication intermediate of hepadnaviruses that forms a minichromosome. In addition, the CTD is required for pgRNA packaging, and core protein also plays an active role in initiating reverse transcription and in mature nucleocapsid envelopment.

Polymerase/Reverse Transcriptase

Not long after the identification of an HBV-like virus in ducks, the DHBV model was used to demonstrate that DHBV genome replication utilizes an RNA intermediate, implying that hepadnaviruses replicate via reverse transcription. While reverse transcription is a mechanism employed by many viruses, hepadnaviruses approach genome replication with a number of unique features. The 90 kD, 838aa polymerase protein of HBV (reverse transcriptase/RT/Pol/P) is made up of 3 functional domains and a variable spacer region. At the N-terminus is the terminal protein (TP) domain, which is important for multiple facets of the initiation of genome replication. This region, despite its important role in P binding to the pgRNA, RNA packaging, and protein-priming, is a unique domain that is not shared by any non-hepadnavirus RTs. A variable spacer separates the TP domain from the RT domain, and studies have shown that nearly all aa within the variable spacer region can be mutated without altering P function. In fact, only 3 cysteine residues within the C-terminal end of the spacer region, along with a fourth in the N-terminal side of the RT domain, are thought to be important for RT/pol function.

The RT domain is responsible for genome replication by reverse transcribing the pgRNA to form the (-)-strand of the DNA genome and subsequent use of the (-)-strand as a template for synthesis of the (+)-strand of the DNA. This domain shares significant homology to the RT of other retroviruses. The RT domain is the only current anti-HBV therapeutic target, which is based on the efficacy of nucleoside analogs to inhibit the human immunodeficiency virus (HIV) RT. In fact, portions of HBV RT can be replaced by homologous portions of HIV RT; this can generate an active RT that can function to produce mature HBV virions.

The final domain, P, is the RNase H domain. This domain is responsible for degrading the pgRNA template during synthesis of the (-)-strand of the DNA genome. Coordination of metal ion binding, which is important for RNase H activity, is achieved through 4 conserved carboxylates. Studies of the RNase H domain have also shown that purified recombinant RNase H domain is functional in vitro and that the RNase H domain of P is important for pgRNA packaging.

X Protein

HBx is the only regulatory protein encoded by HBV. It is a 154aa, 17 kD protein that is encoded by the smallest HBV ORF. Various studies have provided considerable evidence that HBx plays an essential role during HBV replication. Specifically, studies have shown that HBx is bound to cccDNA, that HBx is required for transcription from cccDNA, and that downstream HBx-mediated effects are required for HBV replication. Importantly, studies of other mammalian hepadnaviruses have also supported the role of their respective X proteins in viral replication. For example, two different studies demonstrated that the WHV X protein is required for WHV replication in vivo, although another study did show a low level of viral replication from a WHx-deficient WHV mutant in infected woodchucks. Similarly, viral replication was detected from tg mice expressing either wild-type or an HBx-null HBV mutant; however, when the HBx-null mice were crossed with HBx-tg mice, levels of HBV replication surpassed those seen in wild-type HBV-tg mice. A similar experiment using hydrodynamic tail vein injection of a plasmid encoding either the wild type HBV genome or an HBx-deficient mutant HBV showed a significant decrease in the levels of HBV replication in the absence of HBx, which could be restored through co-injection of the HBx-deficient mutant HBV and a plasmid encoding HBx. This indicates that while HBx may not be absolutely required for HBV replication in these systems, it undoubtedly enhances the levels of replication. Moreover, studies of direct HBV infection of mice with humanized livers demonstrated that only infection with wild type HBV, and not HBx-deficient virus, could result in HBV replication. A similar requirement for HBx in HBV replication has been shown in human HepG2 hepatoblastoma cells and in primary rat hepatocytes. Importantly, the requirement for HBx was also confirmed in primary human hepatocytes directly infected with either wild-type or HBx-deficient HBV.

The lack of a single accepted model for studies of HBV and HBx has created some confusion about the overall consequences of HBx expression for HBV replication and hepatocyte physiology. HBx-related studies have often been performed in transformed or immortalized cell lines and with different levels of HBx expression, leaving the impact of HBx on a normal hepatocyte incompletely understood. While HBx is generally considered to have oncogenic potential, it is yet to be determined if it is directly oncogenic or simply acts as a co-factor in HCC development, as both effects have been demonstrated in different HBx-tg mouse models. It is important to recognize that a strongly oncogenic HBx would not be consistent with the biology of HBV-associated HCC, which involves decades of a chronic HBV infection, and it is more likely that HBx plays a cofactor role in the development of HBV-induced liver cancer. The hypothesis that HBx-induced subtle changes in hepatocyte physiology sensitize cells to other oncogenic signals, while facilitating HBV replication, is more consistent with the biology of HBV-associated HCC. Peripheral evidence for the oncogenic potential of HBx comes from the fact that hepadnavirus-associated HCC seems to be restricted to mammalian hepadnaviruses, which each express a form of the X protein. Avian hepadnaviruses, which do develop a chronic infection but do not cause HCC, either do not express an X protein or express a highly divergent form.

HBx is a multifunctional protein that can modulate many hepatocyte signaling cascades and factors that have also been linked to mechanisms that underlie cellular transformation. For example, HBx can modulate calcium, apoptosis, and proliferation signals, among other pathways, and can activate numerous transcription factors, including activator proteins 1 and 2 (AP-1 and AP-2), nuclear factor of activated T cells (NFAT), and nuclear factor kappa-light-chain-enhancer of activated B cells (NF-κB). HBx can also regulate cellular signaling factors, such as Wnt/β-catenin, p53, and Akt, that have been implicated in HCC. Recently, modulation of miRNA expression has also been included in the functions of HBx. It is possible that the many functions attributed to HBx could actually be the result of a few fundamental upstream HBx functions that can affect multiple downstream cellular signal-transduction pathways in a context-dependent manner. Interestingly, while HBV replication in established HBV-associated HCCs is typically absent, a number of groups have shown that these tumors can still express HBx from fragments of the HBV genome that have integrated into the host genome. The presence of HBx in these cells could mean that HBx might be active in these HCC cells, even in the absence of replicating HBV, and potentially contribute to HCC development or maintenance.

HBV Life Cycle

Studies have shown that the species specificity and hepatotropic nature of HBV are due to at least two different layers of cellular factors. The first is the hepatocyte-specific expression of the recently described HBV receptor, human sodium taurocholate cotransporting peptide (hNTCP/SLC10A1). hNTCP is only expressed on human hepatocytes, and mouse NTCP cannot be bound by HBV, which correlates with the inability of HBV to directly infect mouse hepatocytes. The second level of cell-specificity of an HBV infection is controlled by hepatocyte-specific transcription factors such as HNF1α and

HNF4α; these control post-entry, downstream stages of the HBV life cycle. Evidence for the additional role of intracellular factors for controlling the cell-specificity of an HBV infection comes from the observation that humanized-mouse NTCP, in which the binding residues from mouse NTCP are replaced by hNTCP, allows binding of HBV to the receptor but does not result in a productive HBV infection when expressed in mouse cells. Studies using hepatitis D virus (HDV), which is a satellite virus requiring HBV envelope proteins for entry into a cell, demonstrated that the 75 aa at the N-terminal portion of the PreS1 domain of L-HBsAg are required residues responsible for binding to the viral receptor. In addition, it was shown that N-myristylation of the PreS1 domain is required for infectivity, but not HBV virion assembly. In fact, a myristylated peptide consisting of only the first 47 aa of the preS1 domain is able to bind to hNTCP and inhibit the binding of HBV. Additional studies have suggested a role for heparin sulfate proteoglycans in the initial stages of HBV binding to hepatocytes, including the recent identification, using an RNAi-based screen in Huh7 cells stably expressing hNTCP, of glypican 5 as an HBV and HDV entry factor.

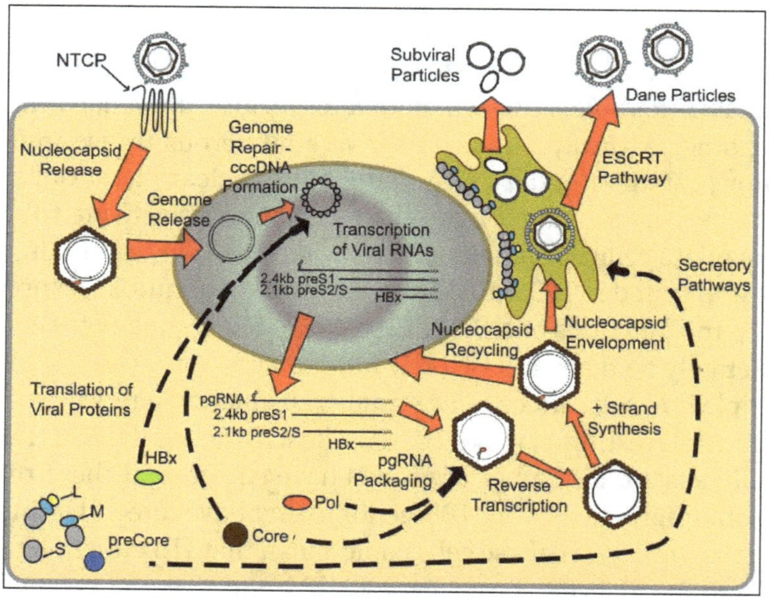

Life cycle of hepatitis B virus (HBV): Mature HBV virions enter hepatocytes through the sodium taurocholate cotransporting polypeptide receptor on the cell membrane. After release from the viral envelope, the nucleocapsid is then transported to the nucleus where the genome is repaired to form covalently-closed circular DNA (cccDNA). Using cccDNA as the template, viral RNAs are transcribed and exported into the cytoplasm where they are translated to form the viral proteins. Additionally, pregenomic RNA (pgRNA) is packaged by core protein, along with the polymerase protein, and the viral genome is replicated through reverse transcription of the pgRNA to form the - strand, followed by partial synthesis of the + strand. Mature nucleocapsids can then either be recycled back to the nucleus to maintain a pool of cccDNA, or enveloped and secreted through the ESCRT pathway.

Although amino acid sequences of both preS1 and hNTCP that affect binding of HBV to hNTCP are known, the lack of an effective model system that mimics a robust natural infection has hampered a complete understanding of aspects of the HBV life cycle immediately following receptor-binding. The observation that preS1 binds to clathrin heavy chain and the adapter protein AP-2 in immortalized primary human hepatocytes, and that knockdown of these proteins inhibits infection, suggests that the HBV-hNTCP complex may enter the cell through clathrin-mediated endocytosis. Once in the cell, the HBV DNA is delivered into the nucleus by mechanisms that remain unclear. One potential mechanism is the active transport of the nucleocapsid through nuclear pores. Another potential mechanism involves CTD phosphorylation of the core protein, which is thought to expose nuclear localization signals, leading to nucleocapsid disintegration at the nuclear pore and transfer of the polymerase-bound, relaxed circular (rc) form of the HBV DNA into the nucleus. The single-stranded gaps in the rcDNA are repaired either through (+)-strand extension by the HBV polymerase or through repair activity of host proteins, and cccDNA is formed as a nucleosome-bound minichromosome in the nucleus. The observations that some HBV-tg mice do not produce cccDNA, and that nucleoside analogues that inhibit the RT function of polymerase do not prevent cccDNA formation, suggest that the production of cccDNA likely involves specific host factors. In addition to studies suggesting a role for cellular histones in cccDNA formation, evidence also exists showing that cccDNA is bound to both core protein and HBx and that this influences the structural arrangement of the cccDNA episome and the epigenetic regulation of cccDNA. Although multiple studies have suggested that HBx is not required for cccDNA formation, transcription of viral RNA from cccDNA is lost in the absence of HBx, and HBx has been suggested to regulate levels of cccDNA histone acetylation and methylation. Host RNA polymerase II uses cell-specific transcription factors and cccDNA, which serves as the template for all viral transcripts, to produce 5'-capped and 3'-polyadenylated RNA transcripts. Translation of the viral transcripts occurs in the cytoplasm following nuclear export.

While a portion of the pgRNA is translated, forming the pool of core and polymerase proteins, pgRNA also serves as the template for reverse transcription. This requires encapsidation of pgRNA by 120 dimers of core protein to form the nucleocapsid. This occurs through a complex cascade of events involving multiple viral and host proteins. Specifically, the 5' end of the pgRNA contains an encapsidation signal, termed ε, which is recognized and bound by polymerase. Studies have also shown that the 5' cap structure is required for packaging of the pgRNA; however, polyadenylation is not required. In addition, interaction of pgRNA-bound polymerase with the 5' cap and host eIF4E leads to encapsidation of this entire RNP complex, resulting in cellular eIF4E within the viral nucleocapsid. Cellular heat shock proteins have also been suggested to play a similar role in stabilizing the binding of polymerase to ε.

Once packaged, reverse transcription is initiated through priming by the polymerase from a specific tyrosine residue within the N-terminal, TP domain of the polymerase. A bulge region within ε supplies the template for the first 3-4nt of the (-)-DNA strand before translocation to a matching acceptor motif in the 3' direct repeat 1* (DR1*). This strand is then extended until completion, resulting in a unit length (-)-DNA strand copy of the pgRNA that contains an additional 10nt terminal redundancy (r). The majority of pgRNA is degraded during DNA synthesis by the RNase H activity of polymerase, with the remaining bases serving as the 5' primer for synthesis of the (+)-strand. Direct extension of this primer from its 5' position results in a double-stranded linear form of the genome that is replication incompetent. This double-stranded linear form does, however, seem to play a role as the main form of HBV DNA that can be integrated into the host genome. Instead of direct extension of the RNA primer from its 5' location, successful rcDNA formation can occur only after the RNA primer is translocated to the 3' DR2 sequence. Once on DR2, the RNA primer is extended towards the 5' end of the (-)-DNA strand. Because r on the other end has the same sequence, exchange of the two ends allows (+)-strand synthesis to continue. As with the previous translocations, additional cis-acting elements are likely playing a role in long-range base-pairing, which allows the close juxtaposition of these donor and acceptor sites that can otherwise be separated by kilobases of sequence. In addition, recent evidence has suggested a role for core protein in regulating DNA synthesis, as mutations in core protein inhibit the synthesis of the (+)-strand of DNA.

Replication occurs in core particles in the cytosol of an HBV-infected hepatocyte, and the final product of DNA synthesis is the encapsidated, partially double-stranded rcDNA with the polymerase still bound to the 5' end of the (-)-DNA strand. This nucleocapsid can then proceed in one of two directions. The first is shuttling of the nucleocapsid back to the nucleus to amplify and maintain a stable pool of cccDNA. The levels of envelope proteins influence this recycling, with decreased amounts of HBsAg promoting shuttling of the nucleocapsid to the nucleus. In particular, levels of L-HBsAg directly influence shuttling back to the nucleus, and these findings correlate well with the early establishment of a cccDNA pool, followed by identification of secreted infectious HBV. The result is a pool of cccDNA that contains a fluctuating number of copies (typically less than 10) of cccDNA per cell, which can be maintained in the cell for years. Additionally, it has been suggested the half-life of a single cccDNA molecule is between 33 and 57 days, underscoring the role of cccDNA in maintaining HBV persistence.

The second potential HBV nucleocapsid-associated process is envelopment by HBV envelope glycoproteins residing in the ER membrane. Interestingly, mechanisms exist that may limit envelopment of capsids containing immature HBV genome; however, these mechanisms remain incompletely understood. For example, it has been suggested that only mature rcDNA-containing nucleocapsids are enveloped, while ssDNA or RNA containing nucleocapsids are not secreted from the cell. Studies utilizing an RNase H-deficient polymerase, which renders the virus unable to initiate (+)-DNA strand synthesis, have suggested that only completion of the (-)-DNA strand is required for envelopment, and specific mutations in core protein can allow envelopment of immature nucleocapsids. The mechanisms associated with this selectivity are unknown, although the phosphorylation state of core protein, likely influenced by the nucleic acid species inside the capsid, could be playing a role. Specifically, studies have shown that core protein isolated from DNA-containing capsids is dephosphorylated (after the prior phosphorylation required for pgRNA packaging and reverse transcription) in a specific C-terminal region, while immature nucleocapsids remain phosphorylated at at least 6 different sites. The overall secretion of infectious HBV Dane particles has been hypothesized to be as little as 1-10 virions per cell per day, which, because of the large number of cells in the liver, can account for high in vivo HBV titers, but can hinder in vitro research requiring isolation of large amounts of infectious virus. Secretion of Dane particles was originally thought to follow the same secretory pathway as the much more abundant SVP, with the envelope proteins residing within the ER-golgi intermediate compartment from where they could bind the DNA-containing capsid, enter the lumen, and be secreted from the cell. Recent evidence has suggested, however, that mature HBV virions are secreted from the cell using a pathway that is dependent on proteins involved in the endosomal sorting complex required for transport (ESCRT) pathway, which forms multivesicular bodies. One characteristic that needs to be considered regardless of the pathway of HBV secretion is the seemingly contradictory conformations of L-HBsAg, with the domains required for both interaction with the nucleocapsid and hNTCP being needed on opposite sides of the membrane. This is addressed by the fact that nearly half of L-HBsAg changes transmembrane conformation after translation, to expose the preS domains within the ER lumen.

Model Systems used in the Study of HBV

Each member of the hepadnavirus family has a narrow host range that is thought to be defined primarily by the interaction between the virus and a specific cell-surface receptor that is present on host hepatocytes. Available cell culture systems for studying the life cycle of the Hepadnaviridae are limited. Typically, members of a hepadnavirus family can only directly infect hepatocytes within the liver of their respective avian or mammalian hosts or cultured, well-differentiated primary hepatocytes that are derived from these hosts. This has hampered the capabilities of researchers to study a natural HBV infection.

Due to the limited host range of HBV, few suitable animal models exist for studying an in vivo HBV infection. Closely related viruses, such as DHBV and WHV, have been used in their respective host animals as surrogate models for understanding overall hepadnavirus biology. These studies have been instrumental in establishing our understanding of the viral life cycle, including the identification of DNA replication through an RNA intermediate, the establishment of a pool of cccDNA as a mechanism for maintaining chronic infection, and the course of both acute and chronic infection.

The treeshrew, Tupaia belangeri, is a small animal model and is one of the very few animals that can be experimentally infected with HBV. Genomic analysis has placed the treeshrew phylogenetically between humans and rodents, and this similarity to primates has spurred its use as a model for a broad range of studies, including as a model for viral hepatitis. Specifically, Tupaia belangeri has been used as a model to study the immediate effects of HBV infection on gene expression in the liver and to identify genes potentially contributing to the development of HBV-associated HCC. In fact, freshly isolated primary treeshrew hepatocytes were recently used in multiple studies in which hNTCP was identified as the HBV receptor. Recent studies also suggest that neonatal exposure of treeshews to HBV can lead to a disease progression similar to what is seen in humans, with development of a chronic infection leading to the eventual development of HCC. Unfortunately, a relatively low HBV infection efficiency and lack of genetically uniform tree shrew strains has limited their use.

The chimpanzee is the only non-human primate model for HBV infection and, along with the tree shrew, represents one of the only animal models that can be directly infected with HBV. HBV can establish both acuteand chronic infectionsin chimpanzees, and this model has been used most often for modeling the immune response to HBV and the interaction between the virus and host. Studies in chimpanzees have helped to establish the relationship between the innate and adaptive immune response to HBV infection, demonstrating minimal early activation of innate immune mediators and a reliance on CD8+ T cells for viral clearance through interferon γ- and tumor necrosis factor α-dependent mechanisms, in agreement with previous work in HBV-tg mice. Another important use of the chimpanzee model has been as a surrogate model for preclinical drug and vaccine testing. The ethical issues and high costs associated with non-human primate use, however, have limited the use of this model and the recent reevaluation in the United States (one of only two countries to allow chimpanzee research) of the need for chimpanzees in preclinical research will likely diminish their future use even further.

A number of mouse models exist for the study of HBV, and have been reviewed more extensively in the literature. Typically these models can be separated into two categories: HBV-/HBx-tg mice, which constitutively express HBV or HBx, respectively, and mice that are delivered the HBV genome or an HBx-expressing plasmid by hydrodynamic tail-vein injection. Although mouse hepatocytes cannot be directly infected with HBV, the use of tail-vein-delivered DNA or HBV-tg mice allows studies of the impact of HBV replication on hepatocyte physiology; HBx-tg mice similarly aid in the study of HBx-mediated effects. While these mouse models are valuable tools, they do have their drawbacks. For example, there is no inflammatory response against HBV in an HBV-tg mouse, which could alter the cellular pathways activated in these models compared to a natural HBV infection. Additionally, because some HBV-tg mice do not produce HBV cccDNA, there is some concern over whether this system accurately mimics all aspect of HBV replication in humans. Despite this, mouse models have been instrumental in determining a number of important aspects of the HBV life cycle, including the requirement for HBx in HBV replication and the oncogenic potential of HBV and HBx.

More recently, two additional mouse models have been described that may greatly enhance our understanding of the HBV life cycle and HBV-associated disease. The first of these systems is the humanized mouse model, in which the majority of the mouse liver is repopulated with either primary human hepatocytes or human induced pluripotent stem cells. The use of these animals represents a significant advancement, as they support direct infection with HBV and can develop a chronic HBV infection, thereby allowing studies of the impact of an HBV infection on the liver in a cellular context more similar to that seen in the human liver. Some of these chimeric mouse models include both a humanized liver and humanized immune cells, offering the unique opportunity to study the human immune response in a small animal model. Although different techniques for the development of the humanized mouse model exist, a number of groups have adapted their use for the study of HBV. These studies cover a broad range of aspects of HBV biology, including studies of the HBV-mediated immune response, investigation of potential HBV therapeutics, and aspects of the HBV life cycle such as particle formation, receptor binding, and cccDNA regulation.

Another recently described mouse model expresses hNTCP to allow receptor binding by HBV. Currently these mice are limited in their utility for studying HBV, as multiple groups have shown that while HBV can bind to hNTCP expressed in mouse cells, the HBV life cycle does not appear to proceed beyond receptor binding. Conversely, HDV utilizes HBV envelope proteins for its envelopment, and hNTCP-expressing mice have been used for the study of HDV infection. Further work with hNTCP-expressing systems may help to determine species-specificity factors that could ultimately lead to the development of an hNTCP-expressing mouse model useful for the study of HBV infection. Together with the humanized-liver model, these mouse models could greatly contribute to our understanding of the early stages of an HBV infection, including entry, HBV genome transport to the nucleus, and genome repair.

The paucity of in vivo models for studying direct HBV infections, and the limited availability of cultured primary human hepatocytes, has lead many researchers to study HBV replication and the activities of HBV-encoded proteins in immortalized or transformed liver cell lines and in cultured primary hepatocytes derived from small-animal models such as rats or mice. Use of these systems necessitates bypassing the initial receptor-mediated infection of the cell by direct transfection of the HBV DNA genome. Although primary hepatocytes derived from small-animal models, namely rat and mouse, cannot be directly infected with HBV, they can support HBV replication and serve as a surrogate model system for studying the effects of HBV replication and HBV proteins on cellular physiology.

Most hepatocyte cell lines that are used in HBV-related studies are tumor-derived and thus are transformed. Since cellular signaling pathways are significantly altered in cancer, cell lines derived from tumors do not recapitulate the physiology of normal hepatocytes. While the results obtained by using transformed cell lines may be valid in a specific cellular context, caution should be exercised in the interpretation of such results because they may not necessarily represent the effects of HBV on cellular physiology in normal, untransformed hepatocytes, the authentic site of an HBV infection.

In addition to tumor-derived cell lines, some cell lines have been specifically derived from HBV-positive tumors. Examples of cell lines isolated from HBV-positive tumors include the PLC/PRF/5 cell line and the Hep3B cell line, which are human HCC-derived cell lines containing multiple copies of HBV DNA integrated into the host DNA. While the PLC/PRF/5 and the Hep3B cells are active in HBsAg production, they do not produce HBV virions and display no indicators of HBV replication, so results of studies using these cell lines require careful interpretation.

In an attempt to establish a system to study the biology of HBV, and specifically HBV replication, the results of numerous studies demonstrated that HBV DNA could be transfected into many different cell lines, including the hepatoblastoma-derived cell line HepG2 and the hepatoma-derived cell line Huh7 and that HBV could replicate efficiently in these cells. Consequently, HepG2 and Huh7 cells expressing HBV DNA are frequently used in the study of HBV biology. While HepG2 cells and Huh7 cells support HBV replication, similar to nearly all existing human liver cell lines, they cannot be directly infected with HBV, partly due to the low expression levels of hNTCP, the functional cellular receptor for HBV. In order to analyze differences between cells with and without replicating HBV, HepG2.2.15 cells have sometimes been compared to HepG2 cells. HepG2.215 cells were originally derived from HepG2 cells and stably express HBV from two integrated head-to-tail dimers of the HBV genome. Results obtained by comparing HepG2.215 cells to the parental HepG2 cells, however, need to be interpreted with caution; because of the continuous passaging of HepG2.215 cells since their development in 1987, dissimilarities beyond the expression of HBV may exist between HepG2.2.15 cells and the parental HepG2 cells. Due to these dissimilarities, phenotypic differences that are observed between HepG2.215 and HepG2 cells might not be an exclusive consequence of replicating HBV.

Together, the use of exogenously delivered HBV DNA into established cell lines, such as HepG2 and Huh7, and the use of cell lines stably expressing HBV, such as the HepG2-derived cell lines HepG2.2.15 and HepAD38, make up the majority of the studies that have been conducted to understand HBV biology. While these cell culture models have proven extremely valuable to study HBV DNA replication, viral assembly, and virion secretion, they have limitations that prevent them from recapitulating all the aspects of an authentic human HBV infection.

Some recent developments have lead to increased optimism for the development of an effective in vitro system to study the complete HBV life cycle. For example, fusion of primary human hepatocytes with hypoxanthine-guanine phosphoribosyltransferase null [HG-PRT(-)] HepG2 cells led to the establishment of the immortal cell line, HepCHLine-4/-7, that may provide a model system for HBV infection. This cell line supports HBV replication and is susceptible to HBV infection when incubated with serum from HBV-positive patients. However, an uncertain genetic stability during maintenance hampers the use of this system. In addition, the HepaRG cell line, a human hepatoma cell line, can also be directly infected with HBV and supports HBV replication. While this cell line is often used in studies of HBV infection, its use is limited by a low HBV infection efficiency of only about 10-20%

and the need to induce differentiation prior to infection, which involves the maintenance of cells in 2% DMSO for 2 weeks before the induction of differentiation.

The recent discovery of hNTCP as the functional HBV receptor has important implications for basic HBV research and antiviral development. In particular, identification of hNTCP has opened new avenues for the establishment of novel cell culture model systems that can be utilized to understand the effects of natural HBV infection. HepG2 cells expressing hNTCP (HepG2-hNTCP), theoretically, provide a convenient in vitro system for HBV infection. The exogenous expression of hNTCP in HepG2 cells does render them susceptible to HBV infection; however, low levels of infection, typically around 10%, and a requirement for large viral inoculums limit their use. Infection-based studies are hampered even further by the low levels of virus released by HBV-infected cells, believed to be around 1 virion per day, making it difficult to produce the large quantities of virus needed for these types of studies. Despite these issues, HepG2-hNTCP cells provide a valuable in vitro model system for elucidating the effects of natural HBV infection, investigating the complete HBV life cycle including the early steps of an HBV infection, and identifying novel therapeutic options.

As the natural target of an HBV infection, primary human hepatocytes would be the ideal in vitro system for studies of HBV. Unfortunately, cultured primary human hepatocytes lose susceptibility to HBV infection within days of isolation and culture, potentially because hNTCP expression rapidly decreases over time in culture. Interestingly, Rice and colleagues recently reported that induced pluripotent stem cell-derived hepatocytes(iPSC-derived iHeps) can support HBV infection, opening potential new avenues to study HBV biology and virus-host interaction and to test antiviral candidates. However, a long induction process involving differentiation of the iPSCs is required prior to HBV infection, and viral markers of infection can only be detected more than a week after inoculation.

Although studies in immortalized or transformed cells have served as powerful models for studying various aspects of HBV replication and the functions of HBV-encoded proteins, these studies have also demonstrated that HBV-mediated activities, particularly those associated with HBV proteins such as HBx, may vary in different cellular contexts. Alternatively, studies in cultured primary hepatocytes have begun to clarify HBV replication strategies and the function of HBV proteins in a more relevant context. Recently, cultured primary rat hepatocytes have been used to study HBV replication and functions of the HBx protein; HBx activities in cultured primary rat hepatocytes were similar to HBx activities in cultured primary human hepatocytes, supporting the use of cultured primary rat hepatocytes as a model system for studying the impact of HBV on hepatocyte physiology.

HBV and HCC

HCC, which accounts for 80-90% of all liver cancers, is one of the most common and most deadly cancers worldwide. Globally, liver cancer is the sixth most common and second deadliest cancer, with an incidence to mortality rate near 1. Epidemiological studies have identified chronic HBV infection of the liver as the leading risk factor for HCC development. Despite the availability of a vaccine, 350-500 million people worldwide are chronically infected with HBV and, depending on age and route of infection, as many as 25% of these individuals could go on to develop HBV-associated HCC. The number of cases of HCC that are attributed to HBV will likely increase as occult infections become both better reported and better understood.

The molecular mechanisms that link a chronic HBV infection to HCC development are incompletely understood but are likely subtle considering that HBV-associated HCC usually occurs in the context of a decades-long chronic HBV infection. Studies have focused on three main factors that may be involved in the development of HBV-associated HCC: chronic inflammation accompanied by destruction and regeneration of hepatocytes, consequences of HBV DNA integration into host genome, and the potential effects of HBV proteins such as HBx.

One particularly important intermediate aspect of a decades-long chronic HBV infection includes the development of HBV-associated cirrhosis prior to HCC development. It is generally accepted that the majority, potentially as much as 70-90%, of all HCC occurs in the context of decompensated cirrhosis, and a strong relationship exists between chronic HBV infection and cirrhosis. In fact, a recent cohort study demonstrated that the cumulative lifetime risk of developing HBV-associated cirrhosis is 41.5% for chronically infected patients, with a cumulative risk of developing HCC of 21.7%. Therefore, establishing a clearer understanding of the cellular mechanisms associated with the intermediate stages of chronic disease, particularly development of cirrhosis, could enhance the overall understanding of causes of HBV-associated HCC.

Numerous aspects of an HBV infection could be playing a role in the development of HCC. It is logical to assume that hepatotropic viruses such as HBV, which alter hepatocyte physiology as part of, or a consequence of their replication, may disrupt normal hepatocyte and overall liver functions. Many of these disruptions and alterations, either through viral replication or activities of viral proteins such as HBx, could be playing a role in the development of downstream HBV-associated HCC. For example, HBV has been shown to disrupt cell cycle regulation, alter apoptotic pathways, alter hepatocyte metabolism and alter miRNA expression and miRNA-mediated regulation. Many of these studies have focused on multiple cellular signal transduction pathways, including those involving Ras/mitogen-associated protein kinases (MAPK), mechanistic target of rapamycin (mTOR), PI3K/Akt and NFκB, among many others. Each of these pathways and factors, while also important for HBV replication, are main mediators of hepatocyte functions. As such, disruption can have a major impact on hepatocyte physiology, which has generated a considerable amount of interest in their potential role as mechanisms for the development of HBV-associated HCC.

HBV and the Cell Cycle

As with many viruses, HBV must optimize the cellular environment for viral replication. In the case of HBV, this involves inducing hepatocytes to exit quiescence and enter into an active cell cycle, and the status of cell proliferation pathways can have a significant impact on HBV replication. For example, in primary rat hepatocytes, HBV moved cells from G0 into and through G1, but stalled progression before the hepatocytes were able to reach S phase, and this regulation of the cell cycle is required for HBV replication in primary rat hepatocytes. Studies in cell lines suggest a similar HBV-mediated regulation of cell cycle progression, with HBV stalling progression of the cell cycle before entry into S phase in both Huh7 cells expressing HBV and the HBV-expressing HepG2.2.15 cells. Interestingly, studies have also shown decreased proliferation of HepG2.2.15 cells, in comparison to HepG2 cells, along with HBV-mediated alteration of cell cycle regulators leading to a G1 phase arrest. Another study, however, in Huh7 cells and primary marmoset hepatocytes, demonstrated an HBV-mediated stall in the G2 phase of the cell cycle. While somewhat contradictory, these results together correlate well with the results of other studies showing that HBV replication is increased when the cell cycle is arrested in either G1 or G2, but HBV replication is decreased during S phase, when cellular DNA synthesis would be higher, potentially depleting the pool of nucleotides that would be available for HBV replication.

Much of the HBV-mediated regulation of the cell cycle appears to be through the activity of the HBx protein. Multiple studies in primary hepatocytes have demonstrated that HBx alters cell cycle regulators, including decreasing p15 and p16 expression, decreasing DNA synthesis, and increasing p21, p27, cyclin D1, and cyclin E expression. Together these results suggest that HBx drives hepatocytes into the cell cycle but increases expression of inhibitors that prevent progression beyond G1. This HBx-mediated regulation of the cell cycle could have a long-term impact on hepatocyte physiology, altering hepatocyte proliferation pathways and contributing to the development of HBV-associated disease and HCC.

HBV and Metabolism

Because of the primary role of the liver as a metabolic organ, a growing body of research has begun to investigate the impact of HBV infection on metabolic pathways in HBV-infected cells. In fact, HBV has been referred to as a "metabolovirus" due to the perceived intersection between HBV gene expression and control of cellular metabolism. Specifically, a number of groups have examined the role of HBV in lipid metabolism, especially considering the well-established link between hepatocyte lipids and various stages of the HCV life cycle and the recent identification of a bile salt transporter as a functional receptor for HBV.

The influence of HBV infection on hepatocyte metabolism was recently brought to the forefront with the identification of hNTCP, the primary bile salt transporter in hepatocytes, as a functional HBV receptor. Interestingly, the binding of HBV, specifically the preS1 domain of L-HBsAg, to hNTCP directly interferes with the normal function of hNTCP suggesting competition for binding motifs within the receptor. Furthermore, point mutations in hNTCP that abolished binding of preS1 also blocked the ability of the receptor to bind taurocholate, suggesting that by binding to hNTCP, HBV could dramatically alter hepatic bile acid uptake.

Recently, HBV-mediated inhibition of normal hNTCP function was extended further using a biochemical profiling approach in which human liver chimeric mice were infected with HBV, and the impact on cholesterol metabolism was determined. Interestingly, similar results were seen in a comparison of HBV-infected humanized mice, mice treated with the HBV entry inhibitor Myrcludex-B, which mimics the preS1 domain and binds to hNTCP to block HBV infection, and liver biopsy samples from chronically HBV-infected individuals. Together these results indicate that the binding of HBV to hNTCP inhibits bile acid uptake, which stimulates bile acid synthesis pathways. One interesting caveat to these studies is the relatively novel use of a direct infection system, which drastically alters the question being asked in the experiment. For example, by also using the preS1 mimic Myrcludex-B, the studies are specifically addressing the impact of receptor binding by HBV, and not the cellular impact of HBV replication. This is in contrast to some previous work, which has typically been done using systems that bypass the infection step. An example of the importance of this distinction is that while the study using HBV-infected human-liver chimeric mice demonstrated that by binding to hNTCP, HBV alters the levels of nuclear farnesoid X receptor (FXR) and small heterodimer partner (SHP), previous work in an HBV-tg mouse model (which bypasses the infection step, among other differences) showed that depletion of FXR and SHP signaling did not diminish viral replication or transcription. This suggests that although HBV binding to its receptor alters the expression of these transcription factors, this alteration might not affect HBV replication. Further research would be needed to determine the relevance of similar contrasting results within different model systems.

In addition to the functional inhibition of a major bile salt transporter, evidence from other studies has suggested that HBV replication may be intimately associated with central metabolic pathways. For example, multiple transcription factors associated with hepatic metabolic processes, including HNFs, peroxisome proliferator-activated receptors (PPARs), and FXRcan all be recruited to the HBV genome. Moreover, studies in vitro have shown that exogenous addition of bile acids to HBV-expressing cells can increase HBV replication.

Induction of gluconeogenesis enhances HBV replication, and HBx has been shown to increase expression of multiple gluconeogenic genes, potentially contributing to the central role of HBx in HBV replication. Recent RNA-seq analyses of HBV-expressing Huh7 cells and primary rat hepatocytes also detected decreased expression of GLUT2, the main hepatic glucose transporter. Investigation of the effect of fasting glucose levels on HBV replication revealed a link, albeit minor, between the metabolic state of the cell and the level of HBV replication, and both gluconeogenesis and lipogenesis are under the same transcription factor control as HBV replication. Studies have linked metabolic changes to effects of HBV proteins. For example, some of the earliest functions attributed to HBx involved its regulation of metabolic pathways, such as HBx-mediated activation of the Ras-Raf-MAPK pathway, a central pathway involved in the response to nearly all changes that affect metabolism. Protein kinase C (PI3K) is also activated by HBx, which correlates with recent results suggesting HBV and HBx activate the PI3K/Akt pathway, reducing HBV replication. In addition, studies have shown that mutant L-HBsAg can activate mTOR signaling, ultimately causing increased lipogenesis.

When considered in combination, these studies support the characterization of HBV as a "metabolovirus", and HBV responds to and causes significant metabolic changes in the cell. While the clinical impact of this altered metabolic regulation remains unknown, some studies have suggested that HBV can "help" reduce fatty liver disease. The overall link between HBV and metabolic syndrome remains less clear; however, studies that consider the impact of direct infection through a major metabolic receptor may help to enhance our understanding of the link between HBV and metabolic pathways, and how this relationship may impact the metabolic state of the liver and the development of HBV-associated disease and HCC.

HBV and Apoptosis

Despite the many studies that have investigated the effect of an HBV infection and expression of HBV proteins on hepatocyte pro- and anti-apoptotic signaling pathways, the interplay of an HBV infection and hepatocyte apoptotic signaling pathways remains incompletely understood. Because an HBV infection is non-cytopathic, it would be expected that HBV either inhibits or has little effect on apoptotic pathways. Evidence has suggested, however, an HBV-mediated effect on cellular apoptosis that is cell-type- or cell-context-dependent. Some of these differing effects can be attributed to HBx activities, which often have divergent functions depending on context. In the case of apoptosis, some studies have shown that HBx can inhibit apoptosis or have no effect on apoptosis, while other studies have shown that HBx can activate apoptotic pathways or sensitize cells to pro-apoptotic stimuli. The context-dependent apoptotic effects of HBx were clearly shown by studies demonstrating that HBx sensitized dedifferentiated hepatocytes to apoptosis, while HBx-expressing hepatocytes that remained differentiated were resistant to apoptotic stimuli. This underscores the importance of using relevant cell systems for studying the cellular impact of HBV replication and protein expression on cell physiology. HBx was also shown to have divergent apoptotic functions in the context of HBV replication. Studies in primary hepatocytes demonstrated that HBx can have both a pro- and anti-apoptotic effect, depending on the cellular context of HBx expression. Specifically, inhibition of apoptosis was linked to HBx-mediated activation of NFκB; however, when activation of NFκB was blocked, HBx induced apoptosis through pathways involving the mitochondria permeability transition pore (MPTP), a critical pore that spans the inner mitochondrial and outer mitochondrial membranes and affects numerous mitochondrial functions, including mitochondrial control of apoptosis. Whether HBV, through functions of HBx, regulates apoptosis as a mechanism for regulating viral replication or enhancing viral spread is currently unknown. Although activation of apoptosis may impact both viral spread, and immune evasion, recent evidence suggests that alteration of apoptosis during HBV infection is unlikely to impact viral spread. Due to the regenerative nature of hepatocytes, it is also possible that the impact of HBV on apoptosis may fluctuate during the course of infection, as regenerating hepatocytes have different active signaling pathways than quiescent cells, and these could have differing influences on apoptotic stimuli. Interestingly, both the activation and the inactivation of apoptosis could be playing a role in the long-term development of HBV-associated HCC: enhanced regeneration associated with HBV-mediated activation of apoptosis could lead to selection of apoptosis-resistant cells, while inhibition of apoptosis could lead to unchecked proliferation and the accumulation of transforming mutations. Although the exact mechanisms that underlie HBV and HBx regulation of apoptosis remain incompletely understood, the cellular impact of altered apoptotic signaling could significantly contribute to the downstream development of HBV-associated disease and HCC.

HBV and MicroRNAs

Potentially spurred by the discovery of the required role of miR-122 in successful HCV replication, multiple groups have begun to investigate how cellular miRNAs may impact various aspects of HBV biology, and alternatively, how HBV may impact the expression of cellular miRNAs.

While a wide range of cellular miRNAs has been investigated for their role in regulating or being regulated by HBV, none has been studied as extensively as miR-122. This miRNA, which makes up 50-70% of the total miRNA pool in hepatocytes has been shown to have many different roles in the context of HBV replication. Although conflicting reports do exist, it seems that miR-122 functions as an antiviral miRNA, potentially through multiple mechanisms. These mechanisms include direct targeting of viral RNAs through miR-122 recognition sites in the HBV genome and altered miR-122-mediated regulation of cellular factors involved in regulation of HBV replication, such as heme oxygenase-1, cyclin G1, and pituitary tumor-transforming gene 1 binding factor (PBF). Importantly, multiple studies have also shown an HBV-mediated decrease in the levels of functional miR-122, although the mechanism for this reduction remains unclear. One potential mechanism is a sponge effect, where the HBV transcripts act as a sponge to divert miR-122 away from endogenous targets, although it is unclear whether the levels of HBV transcripts in the cell reach the high levels of target required for this sponge effect to have a biological impact. Interestingly, primary tree shrew hepatocytes, which can be directly infected with human HBV, showed an increase in the levels of miR-122 in response to HBV infection. Further research will be needed to determine if this effect is the result of using a more biologically relevant system, with direct infection, or is an inherent difference between tree shrew and human hepatocytes.

Other miRNA families have also been assessed for their role in HBV replication, including miRNAs with well-established roles as either oncomirs or tumor suppressors. For example, the let-7 family, which function as tumor suppressor miRNAs and are downregulated in multiple cancers including HCC, are decreased in the context of HBV replication, HBx expression, and HBV-associated HCC. The miR-15 family, miR-125 family, miR-17/92 cluster, and miR-199a-3p are all HCC-related miRNAs that multiple groups have studied in the context of HBV. Although the field is still developing and contradictory reports exist, when taken together these reports support the overall impact of HBV on cellular miRNAs. How these miRNAs impact HBV replication, and ultimately HBV-associated disease and development of HBV-associated HCC, remains incompletely understood.

References

- Martinez-Salas E, Saiz M, Sobrino F (2008). "Foot-and-Mouth Disease Virus". Animal Viruses: Molecular Biology. Caister Academic Press. pp. 1–38. ISBN 978-1-904455-22-6
- Dixon; et al. (2008). "African Swine Fever Virus". Animal Viruses: Molecular Biology. Caister Academic Press. ISBN 978-1-904455-22-6
- Weiss AT, Klopfleisch R, Gruber AD (2010). "Prevalence of feline leukaemia provirus DNA in feline lymphomas". J Feline Med Surg. 12 (12): 929–35. doi:10.1016/j.jfms.2010.07.006. PMID 21036089
- Ghanim, M; Czosnek, H (2000). "Tomato Yellow Leaf Curl Geminivirus (TYLCV-Is) Is Transmitted among Whiteflies (Bemisia tabaci) in a Sex-Related Manner". Journal of Virology. 74 (10): 4738–4745. doi:10.1128/jvi.74.10.4738-4745.2000. PMC 111996. PMID 10775612
- Khelifa, M.; Massé, D.; Blanc, S.; Drucker, M. (Jan 2010). "Evaluation of the minimal replication time of Cauliflower mosaic virus in different hosts". Virology. 396 (2): 238–45. doi:10.1016/j.virol.2009.09.032. PMID 19913268
- Hily JM, Scorza R, Malinowski T, Zawadzka B, Ravelonandro M (October 2004). "Stability of gene silencing-based resistance to Plum pox virus in transgenic plum (Prunus domestica L.) under field conditions". Transgenic Res. 13 (5): 427–36. doi:10.1007/s11248-004-8702-3. PMID 15587267
- Wylie, Kristine M.; Weinstock, George M.; Storch, Gregory A. (1 October 2012). "Emerging view of the human virome". Translational Research. 160 (4): 283–290. doi:10.1016/j.trsl.2012.03.006. PMC 3701101. PMID 22683423

Viral Life Cycle 4

- **Viral Entry**
- **Viral Replication**
- **Viral Shedding**
- **Virus Latency**

The life cycle of viruses follow several stages for infecting the host cells. Its stages are attachment, penetration, uncoating, biosynthesis, maturation and release. Its other stages are viral entry, replication, shredding and latency. This chapter closely examines viral life cycle to provide an extensive understanding of the subject.

All viruses depend on cells for reproduction and metabolic processes. By themselves, viruses do not encode for all of the enzymes necessary for viral replication. But within a host cell, a virus can commandeer cellular machinery to produce more viral particles. Bacteriophages replicate only in the cytoplasm, since prokaryotic cells do not have a nucleus or organelles. In eukaryotic cells, most DNA viruses can replicate inside the nucleus, with an exception observed in the large DNA viruses, such as the poxviruses, that can replicate in the cytoplasm. RNA viruses that infect animal cells often replicate in the cytoplasm.

Life Cycle of Viruses with Prokaryote Hosts

The life cycle of bacteriophages has been a good model for understanding how viruses affect the cells they infect, since similar processes have been observed for eukaryotic viruses, which can cause immediate death of the cell or establish a latent or chronic infection. Virulent phages typically lead to the death of the cell through cell lysis. Temperate phages, on the other hand, can become part of a host chromosome and are replicated with the cell genome until such time as they are induced to make newly assembled viruses, or progeny viruses.

Lytic Cycle

During the lytic cycle of virulent phage, the bacteriophage takes over the cell, reproduces new phages, and destroys the cell. T-even phage is a good example of a well-characterized class of virulent phages. There are five stages in the bacteriophage lytic cycle. Attachment is the first stage in the infection process in which the phage interacts with specific bacterial surface receptors (e.g., lipopolysaccharides and OmpC protein on host surfaces). Most phages have a narrow host range and may infect one species of bacteria or one strain within a species. This unique recognition can be exploited for targeted treatment of bacterial infection by phage therapy or for phage typing to identify unique bacterial subspecies or strains. The second stage of infection is entry or penetration. This occurs through contraction of the tail sheath, which acts like a hypodermic needle to inject the viral genome through the cell wall and membrane. The phage head and remaining components remain outside the bacteria.

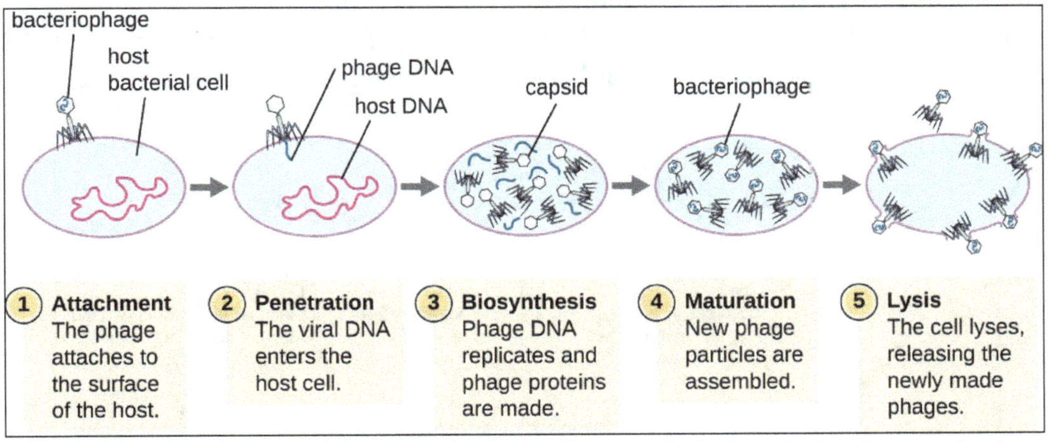

A virulent phage shows only the lytic cycle pictured here. In the lytic cycle, the phage replicates and lyses the host cell.

The third stage of infection is biosynthesis of new viral components. After entering the host cell, the virus synthesizes virus-encoded endonucleases to degrade the bacterial chromosome. It then hijacks the host cell to replicate, transcribe, and translate the necessary viral components (capsomeres, sheath, base plates, tail fibers, and viral enzymes) for the assembly of new viruses. Polymerase genes are usually expressed early in the cycle, while capsid and tail proteins are expressed later. During the maturation phase, new virions are created. To liberate free phages, the bacterial cell wall is disrupted by phage proteins such as holin or lysozyme. The final stage is release. Mature viruses burst out of the host cell in a process called lysis and the progeny viruses are liberated into the environment to infect new cells.

Lysogenic Cycle

In a lysogenic cycle, the phage genome also enters the cell through attachment and penetration. A prime example of a phage with this type of life cycle is the lambda phage. During the lysogenic cycle, instead of killing the host, the phage genome integrates into the bacterial chromosome and becomes part of the host. The integrated phage genome is called a prophage. A bacterial host with a prophage is called a lysogen. The process in which a bacterium is infected by a temperate phage is called lysogeny. It is typical of temperate phages to be latent or inactive within the cell. As the bacterium replicates its chromosome, it also replicates the phage's DNA and passes it on to new daughter cells during reproduction. The presence of the phage may alter the phenotype of the bacterium, since it can bring in extra genes (e.g., toxin genes that can increase bacterial virulence). This change in the host phenotype is called lysogenic conversion or phage conversion. Some bacteria, such as Vibrio cholerae and Clostridium botulinum, are less virulent in the absence of the prophage. The phages infecting these bacteria carry the toxin genes in their genome and enhance the virulence of the host when the toxin genes are expressed. In the case of V. cholera, phage encoded toxin can cause severe diarrhea; in C. botulinum, the toxin can cause paralysis. During lysogeny, the prophage will persist in the host chromosome until induction, which results in the excision of the viral genome from the host chromosome. After induction has occurred the temperate phage can proceed through a lytic cycle and then undergo lysogeny in a newly infected cell.

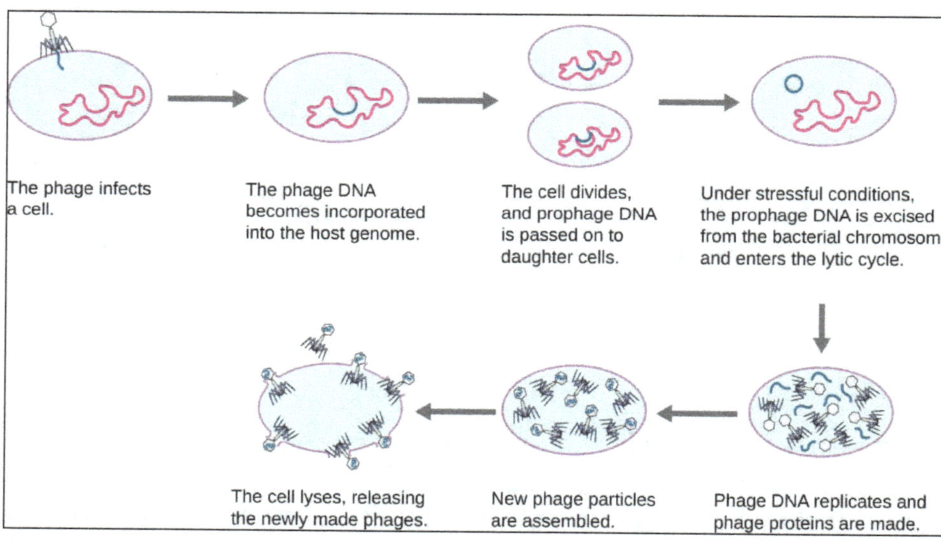

A temperate bacteriophage has both lytic and lysogenic cycles. In the lysogenic cycle, phage DNA is incorporated into the host genome, forming a prophage, which is passed on to subsequent generations of cells. Environmental stressors such as starvation or exposure to toxic chemicals may cause the prophage to be excised and enter the lytic cycle.

Transduction

Transduction occurs when a bacteriophage transfers bacterial DNA from one bacterium to another during sequential infections. There are two types of transduction: generalized and specialized transduction. During the lytic cycle of viral replication, the virus hijacks the host cell, degrades the host chromosome, and makes more viral genomes. As it assembles and packages DNA into the phage head, packaging occasionally makes a mistake. Instead of packaging viral DNA, it takes a random piece of host DNA and inserts it into the capsid. Once released, this virion will then inject the former host's DNA into a newly infected host. The asexual transfer of genetic information can allow for DNA recombination to occur, thus providing the new host with new genes (e.g., an antibiotic-resistance gene, or a sugar-metabolizing gene).

Generalized transduction occurs when a random piece of bacterial chromosomal DNA is transferred by the phage during the lytic cycle. Specialized transduction occurs at the end of the lysogenic cycle, when the prophage is excised and the bacteriophage enters the lytic cycle. Since the phage is integrated into the host genome, the prophage can replicate as part of the host. However, some conditions (e.g., ultraviolet light exposure or chemical exposure) stimulate the prophage to undergo induction, causing the phage to excise from the genome, enter the lytic cycle, and produce new phages to leave host cells. During the process of excision from the host chromosome, a phage may occasionally remove some bacterial DNA near the site of viral integration. The phage and host DNA from one end or both ends of the integration site are packaged within the capsid and are transferred to the new, infected host. Since the DNA transferred by the phage is not randomly packaged but is instead a specific piece of DNA near the site of integration, this mechanism of gene transfer is referred to as specialized transduction. The DNA can then recombine with host chromosome, giving the latter new characteristics. Transduction seems to play an important role in the evolutionary process of bacteria, giving them a mechanism for asexual exchange of genetic information.

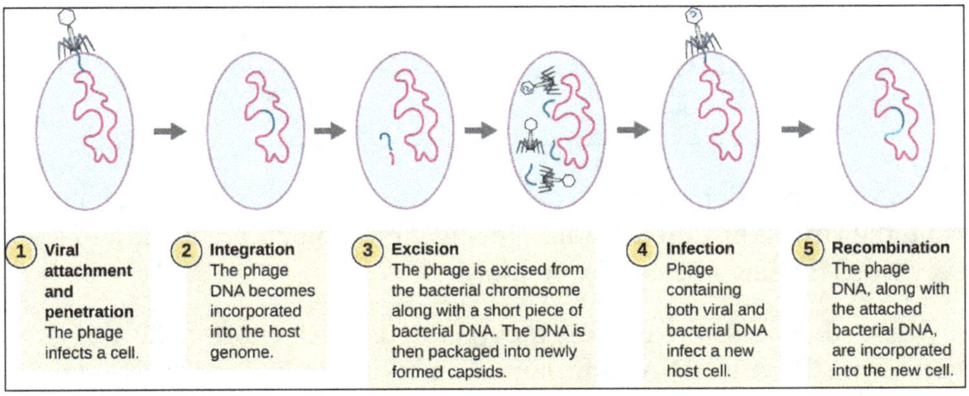

This flowchart illustrates the mechanism of specialized transduction. An integrated phage excises, bringing with it a piece of the DNA adjacent to its insertion point. On re-infection of a new bacterium, the phage DNA integrates along with the genetic material acquired from the previous host.

Life Cycle of Viruses with Animal Hosts

Lytic animal viruses follow similar infection stages to bacteriophages: attachment, penetration, biosynthesis, maturation, and release. However, the mechanisms of penetration, nucleic-acid biosynthesis, and release differ between bacterial and animal viruses. After binding to host receptors, animal viruses enter through endocytosis (engulfment by the host cell) or through membrane fusion (viral envelope with the host cell membrane). Many viruses are host specific, meaning they only infect a certain type of host; and most viruses only infect certain types of cells within tissues. This specificity is called a tissue tropism. Examples of this are demonstrated by the poliovirus, which exhibits tropism for the tissues of the brain and spinal cord, or the influenza virus, which has a primary tropism for the respiratory tract.

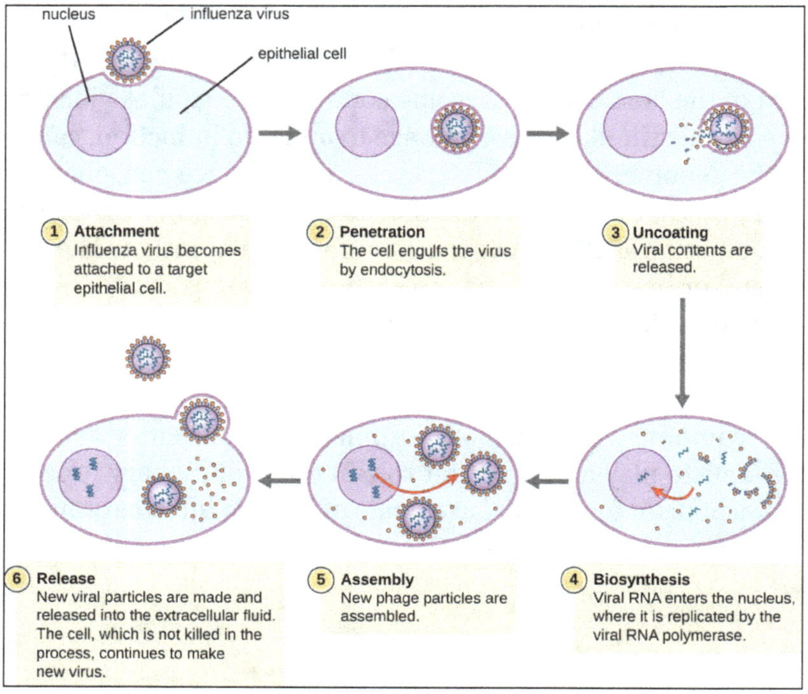

In influenza virus infection, viral glycoproteins attach the virus to a host epithelial cell. As a result, the virus is engulfed. Viral RNA and viral proteins are made and assembled into new virions that are released by budding.

Animal viruses do not always express their genes using the normal flow of genetic information—from DNA to RNA to protein. Some viruses have a dsDNA genome like cellular organisms and can follow the normal flow. However, others may have ssDNA, dsRNA, or ssRNA genomes. The nature of the genome determines how the genome is replicated and expressed as viral proteins. If a genome is ssDNA, host enzymes will be used to synthesize a second strand that is complementary to the genome strand, thus producing dsDNA. The dsDNA can now be replicated, transcribed, and translated similar to host DNA.

If the viral genome is RNA, a different mechanism must be used. There are three types of RNA genome: dsRNA, positive (+) single-strand (+ssRNA) or negative (–) single-strand RNA (–ssRNA). If a virus has a +ssRNA genome, it can be translated directly to make viral proteins. Viral genomic +ssRNA acts like cellular mRNA. However, if a virus contains a –ssRNA genome, the host ribosomes cannot translate it until the

–ssRNA is replicated into +ssRNA by viral RNA-dependent RNA polymerase (RdRP). The RdRP is brought in by the virus and can be used to make +ssRNA from the original –ssRNA genome. The RdRP is also an important enzyme for the replication of dsRNA viruses, because it uses the negative strand of the double-stranded genome as a template to create +ssRNA. The newly synthesized +ssRNA copies can then be translated by cellular ribosomes.

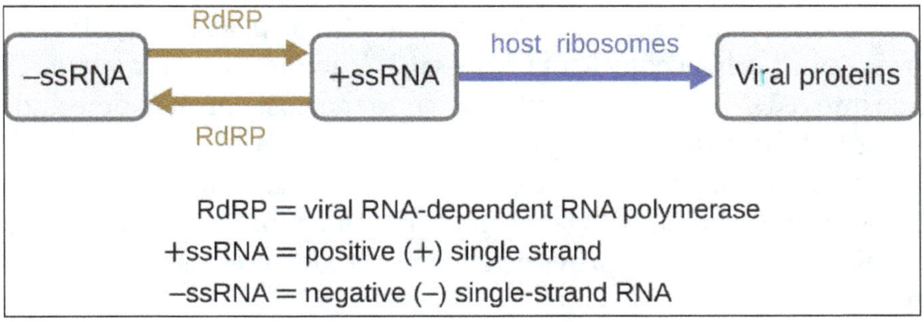

RNA viruses can contain +ssRNA that can be directly read by the ribosomes to synthesize viral proteins. Viruses containing –ssRNA must first use the –ssRNA as a template for the synthesis of +ssRNA before viral proteins can be synthesized.

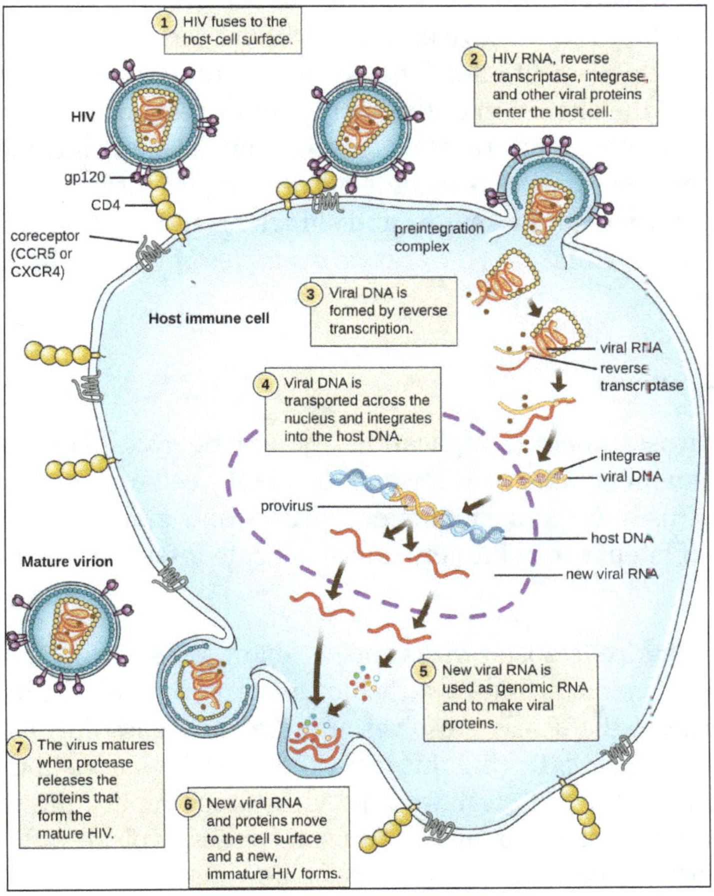

HIV, an enveloped, icosahedral retrovirus, attaches to a cell surface receptor of an immune cell and fuses with the cell membrane. Viral contents are released into the cell, where viral enzymes convert the single-stranded RNA genome into DNA and incorporate it into the host genome.

An alternative mechanism for viral nucleic acid synthesis is observed in the retroviruses, which are +ssRNA viruses. Single-stranded RNA viruses such as HIV carry a special enzyme called reverse transcriptase within the capsid that synthesizes a complementary ssDNA (cDNA) copy using the +ssRNA genome as a template. The ssDNA is then made into dsDNA, which can integrate into the host chromosome and become a permanent part of the host. The integrated viral genome is called a provirus. The virus now can remain in the host for a long time to establish a chronic infection. The provirus stage is similar to the prophage stage in a bacterial infection during the lysogenic cycle. However, unlike prophage, the provirus does not undergo excision after splicing into the genome.

Persistent Infections

Persistent infection occurs when a virus is not completely cleared from the system of the host but stays in certain tissues or organs of the infected person. The virus may remain silent or undergo productive infection without seriously harming or killing the host. Mechanisms of persistent infection may involve the regulation of the viral or host gene expressions or the alteration of the host immune response. The two primary categories of persistent infections are latent infection and chronic infection. Examples of viruses that cause latent infections include herpes simplex virus (oral and genital herpes), varicella-zoster virus (chickenpox and shingles), and Epstein-Barr virus (mononucleosis). Hepatitis C virus and HIV are two examples of viruses that cause long-term chronic infections.

Latent Infection

Not all animal viruses undergo replication by the lytic cycle. There are viruses that are capable of remaining hidden or dormant inside the cell in a process called latency. These types of viruses are known as latent viruses and may cause latent infections. Viruses capable of latency may initially cause an acute infection before becoming dormant.

For example, the varicella-zoster virus infects many cells throughout the body and causes chickenpox, characterized by a rash of blisters covering the skin. About 10 to 12 days postinfection, the disease resolves and the virus goes dormant, living within nerve-cell ganglia for years. During this time, the virus does not kill the nerve cells or continue replicating. It is not clear why the virus stops replicating within the nerve cells and expresses few viral proteins but, in some cases, typically after many years of dormancy, the virus is reactivated and causes a new disease called shingles. Whereas chickenpox affects many areas throughout the body, shingles is a nerve cell-specific disease emerging from the ganglia in which the virus was dormant.

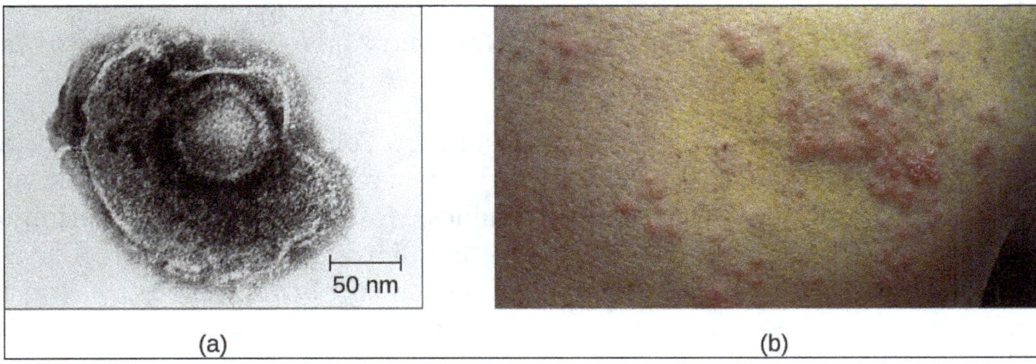

(a) Varicella-zoster, the virus that causes chickenpox, has an enveloped icosahedral capsid visible in this transmission electron micrograph. Its double-stranded DNA genome becomes incorporated in the host DNA. (b) After a period of latency, the virus can reactivate in the form of shingles, usually manifesting as a painful, localized rash on one side of the body.

Latent viruses may remain dormant by existing as circular viral genome molecules outside of the host chromosome. Others become proviruses by integrating into the host genome. During dormancy, viruses do not cause any symptoms of disease and may be difficult to detect. A patient may be unaware that he or she is carrying the virus unless a viral diagnostic test has been performed.

Chronic Infection

A chronic infection is a disease with symptoms that are recurrent or persistent over a long time. Some viral infections can be chronic if the body is unable to eliminate the virus. HIV is an example of a virus that produces a chronic infection, often after a long period of latency. Once a person becomes infected with HIV, the virus can be detected in tissues continuously thereafter, but untreated patients often experience no symptoms for years. However, the virus maintains chronic persistence through several mechanisms that interfere with immune function, including preventing expression of viral antigens on the surface of infected cells, altering immune cells themselves, restricting expression of viral genes, and rapidly changing viral antigens through mutation. Eventually, the damage to the immune system results in progression of the disease leading to acquired immunodeficiency syndrome (AIDS). The various mechanisms that HIV uses to avoid being cleared by the immune system are also used by other chronically infecting viruses, including the hepatitis C virus.

Life Cycle of Viruses with Plant Hosts

Plant viruses are more similar to animal viruses than they are to bacteriophages. Plant viruses may be enveloped or non-enveloped. Like many animal viruses, plant viruses can have either a DNA or RNA genome and be single stranded or double stranded. However, most plant viruses do not have a DNA genome; the majority have a +ssRNA genome, which acts like messenger RNA (mRNA). Only a minority of plant viruses have other types of genomes.

Plant viruses may have a narrow or broad host range. For example, the citrus tristeza virus infects only a few plants of the Citrus genus, whereas the cucumber mosaic virus infects thousands of plants of various plant families. Most plant viruses are transmitted by contact between plants, or by fungi, nematodes, insects, or other arthropods that act as mechanical vectors. However, some viruses can only be transferred by a specific type of insect vector; for example, a particular virus might be transmitted by aphids but not whiteflies. In some cases, viruses may also enter healthy plants through wounds, as might occur due to pruning or weather damage.

Viruses that infect plants are considered biotrophic parasites, which means that they can establish an infection without killing the host, similar to what is observed in the lysogenic life cycles of bacteriophages. Viral infection can be asymptomatic (latent) or can lead to cell death (lytic infection). The life cycle begins with the penetration of the virus into the host cell. Next, the virus is uncoated within the cytoplasm of the cell when the capsid is removed. Depending on the type of nucleic acid, cellular components are used to replicate the viral genome and synthesize viral proteins for assembly of new virions. To establish a systemic infection, the virus must enter a part of the vascular system of the plant, such as the phloem. The time required for systemic infection may vary from a few days to a few weeks depending on the virus, the plant species, and the environmental conditions. The virus life cycle is complete when it is transmitted from an infected plant to a healthy plant.

Viral Growth Curve

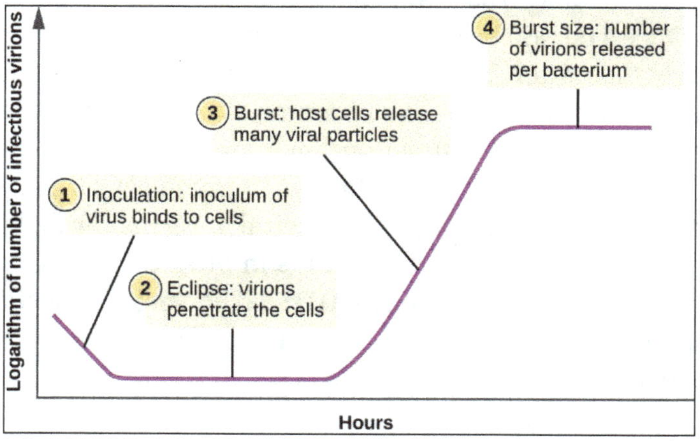

The one-step multiplication curve for a bacteriophage population follows three steps: 1) inoculation, during which the virions attach to host cells; 2) eclipse, during which entry of the viral genome occurs; and 3) burst, when sufficient numbers of new virions are produced and emerge from the host cell. The burst size is the maximum number of virions produced per bacterium.

Unlike the growth curve for a bacterial population, the growth curve for a virus population over its life cycle does not follow a sigmoidal curve. During the initial stage, an inoculum of virus causes infection. In the eclipse phase, viruses bind and penetrate the cells with no virions detected in the medium. The chief difference that next appears in the viral growth curve compared to a bacterial growth curve occurs when virions are released from the lysed host cell at the same time. Such an occurrence is called a burst, and the number of virions per bacterium released is described as the burst size. In a one-step multiplication curve for bacteriophage, the host cells lyse, releasing many viral particles to the medium, which leads to a very steep rise in viral titer (the number of virions per unit volume). If no viable host cells remain, the viral particles begin to degrade during the decline of the culture.

Viral Entry

Viral entry is the earliest stage of infection in the viral life cycle, as the virus comes into contact with the host cell and introduces viral material into the cell. The major steps involved in viral entry are shown below. Despite the variation among viruses, there are several shared generalities concerning viral entry.

Reducing Cellular Proximity

A virus floating around an enclosed space with possible host cells faces a large hurdle, the thermodynamics of diffusion. Because neutrally charged objects do not naturally clump around each other, the virus must find a way to move even near a host cell. It does this by attachment - or adsorption - onto a susceptible cell; a cell which holds a receptor that the virus can bind to. The receptors on the viral envelope effectively become connected to complementary receptors on the cell membrane. This attachment causes the two membranes to remain in mutual proximity, favoring further interactions between surface proteins. This is also the first requisite that must be satisfied before a cell can become infected. Satisfaction of this requisite makes the cell susceptible. Viruses that exhibit this behavior include many enveloped viruses such as HIV and Herpes simplex virus

This basic idea extends to viruses that do not contain an envelope. Well studied examples are the viruses that infect bacteria, known as bacteriophages (or simply phages). Typical phages have long tails used to attach to receptors on the bacterial surface.

Prior to entry, a virus must attach to a host cell. Attachment is achieved when specific proteins on the viral capsid or viral envelope bind to specific proteins called receptor proteins on the cell membrane of the target cell. A virus must now enter the cell, which is covered by a phospholipid bilayer, a cell's natural barrier to the outside world. The process by which this barrier is breached depends upon the virus. Types of entry are:

- Membrane Fusion or Hemifusion State: The cell membrane is punctured and made to further connect with the unfolding viral envelope.
- Endocytosis: The host cell takes in the viral particle through the process of endocytosis, essentially engulfing the virus like it would a food particle.
- Viral Penetration: The viral capsid or genome is injected into the host cell's cytoplasm.

Through the use of green fluorescent protein (GFP), virus entry and infection can be visualized in real-time. Once a virus enters a cell, replication is not immediate and indeed takes some time (seconds to hours).

Entry via Membrane Fusion

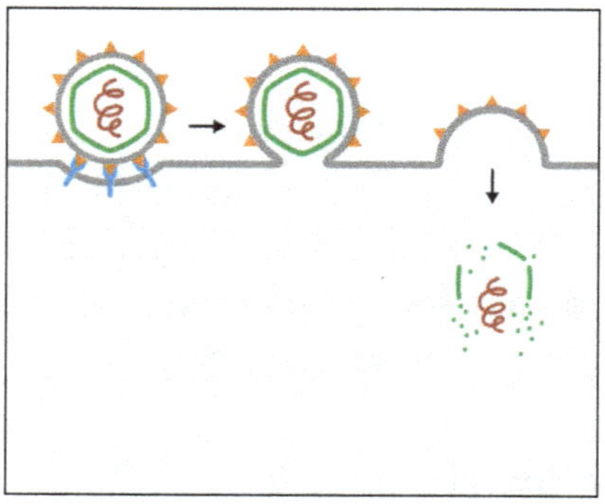

Viral entry via membrane fusion.

The most well-known example is through membrane fusion. In viruses with a viral envelope, viral receptors attach to the receptors on the surface of the cell and secondary receptors may be present to initiate the puncture of the membrane or fusion with the host cell. Following attachment, the viral envelope fuses with the host cell membrane, emptying the now-bare virus into the cell. In essence, the virus's envelope "blends" with the host cell membrane, releasing its contents into the cell. Obviously, this can only be done with viruses that have an envelope (examples of such enveloped viruses include HIV, KSHV and herpes simplex virus.)

Entry via Endocytosis

Viruses with no viral envelope enter the cell through endocytosis; they are ingested by the host cell through the cell membrane. In essence, the virus tricks the cell into thinking that the virus knocking at the door is nothing more than nutrition or harmless goods. A cell, which naturally takes in resources from the environment by attaching goods onto surface receptors and bringing them into the cell, will engulf the virus. Once inside the cell, the virus must now break out of the vesicle by which it was taken up in order to gain access to the cytoplasm. Examples include the poliovirus, Hepatitis C virus and Foot-and-mouth disease virus.

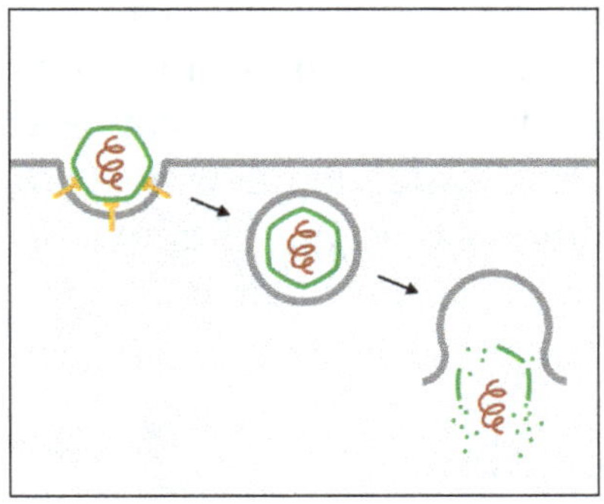

Viral entry via endocytosis.

Many enveloped viruses also enter the cell through endocytosis. Entry via the endosome guarantees low pH and exposure to proteases which are needed to open the viral capsid and release the genetic material inside. Further, endosomes transport the virus through the cell and ensure that no trace of the virus is left on the surface, which could be a substrate for immune recognition.

Entry via Genetic Injection

A third and more specific example, is by simply attaching to the surface of the cell via receptors on the cell, and injecting only its genome into the cell, leaving the rest of the virus on the surface. This is restricted to viruses in which only the gene is required for infection of a cell (most positive-sense, single-stranded RNA viruses because they can be immediately translated) and further restricted to viruses that actually exhibit this behavior. The best studied example includes the bacteriophages; for example, when the tail fibers of the T2 phage land on a cell, its central sheath pierces the cell membrane and the phage injects DNA from the head capsid directly into the cell.

Aftermath

Once a virus is in a cell, it will activate formation of proteins (either by itself or using the host) to gain full control of the host cell, if it is able to. Control mechanisms include the suppression of intrinsic cell defenses, suppression of cell signaling and suppression of host cellular transcription and translation. Often, it is these cytotoxic effects that lead to the death and decline of a cell infected by a virus.

A cell is classified as susceptible to a virus if the virus is able to enter the cell. After the introduction of the viral particle, unpacking of the contents (viral proteins in the tegument and the viral genome via some form of nucleic acid) occurs as preparation of the next stage of viral infection: viral replication.

Viral Replication

Viral replication is the formation of biological viruses during the infection process in the target host cells. Viruses must first get into the cell before viral replication can occur. Through the generation of abundant copies of its genome and packaging these copies, the virus continues infecting new hosts. Replication between viruses is greatly varied and depends on the type of genes involved in them. Most DNA viruses assemble in the nucleus while most RNA viruses develop solely in cytoplasm.

Viral Production/Replication

Viruses multiply only in living cells. The host cell must provide the energy and synthetic machinery and the low molecular-weight precursors for the synthesis of viral proteins and nucleic acids.

The virus replication occurs in seven stages, namely;

- Adsorption,
- Entry,
- Uncoating,
- Transcription/mRNA production,
- Synthesis of virus components,
- Virion assembly,
- Release (Liberation Stage).

Adsorption

It is the first step of viral replication. The virus attaches to the cell membrane of the host cell. It then injects its DNA or RNA into the host to initiate infection. In animal cells these viruses get into the cell through the process of endocytosis which works through fusing of the virus and fusing of the viral envelope with the cell membrane of the animal cell and in plant cell it enters through the process of pinocytosis which works on pinching of the viruses.

Entry

The cell membrane of the host cell invaginates the virus particle, enclosing it in a pinocytotic vacuole. This protects the cell from antibodies like in the case of the HIV virus.

Uncoating

Cell enzymes (from lysosomes) strip off the virus protein coat. This releases or renders accessible the virus nucleic acid or genome.

Transcription/mRNA Production

For some RNA viruses, the infecting RNA produces messenger RNA (mRNA). This is translation of the genome into protein products. For others with negative stranded RNA and DNA, viruses are produced by transcription then translation.

The mRNA is used to instruct the host cell to make virus components. The virus takes advantage of the existing cell structures to replicate itself.

Synthesis of Virus Components

The following components are manufactured by the virus through the host's existing organelles:

- Viral protein synthesis: Virus mRNA is translated on cell ribosomes into two types of virus protein.
- Structural: The proteins which make up the virus particle are manufactured and assembled.
- Non-structural: Not found in particle, mainly enzymes for virus genome replication.
- Viral nucleic acid synthesis (genome replication) new virus genome is synthesized, templates are either the parental genome or with single stranded nucleic acid genomes, newly formed complementary strands. By a virus called polymerate or replicate in some DNA viruses by a cell enzyme. This is done in rapidly dividing cells.

Virion Assembly

A virion is simply an active or intact virus particle. In this stage, newly synthesized genome(nucleic acid), and proteins are assembled to form new virus particles.

This may take place in the cell's nucleus, cytoplasm, or at plasma membrane for most developed viruses.

Release (Liberation Stage)

The viruses, now being mature are released by either sudden rupture of the cell, or gradual extrusion (budding) of enveloped viruses through the cell membrane.

The new viruses may invade or attack other cells, or remain dormant in the cell. In the case of bacterial viruses, the release of progeny virions takes place by lysis of the infected bacterium. However, in the case of animal viruses, release usually occurs without cell lysis.

Lytic Cycle

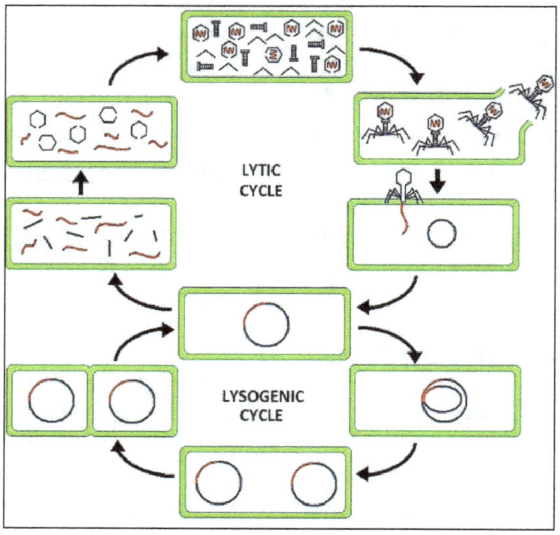

Lytic cycle, compared to lysogenic cycle.

The lytic cycle is one of the two cycles of viral reproduction (referring to bacterial viruses or bacteriophages), the other being the lysogenic cycle. The lytic cycle results in the destruction of the infected cell and its membrane. Bacteriophages that only use the lytic cycle are called virulent phages (in contrast to temperate phages).

In the lytic cycle, the viral DNA exists as a separate free floating molecule within the bacterial cell, and replicates separately from the host bacterial DNA, whereas in the lysogenic cycle, the viral DNA is located within the host DNA. This is the key difference between the lytic and lysogenic (bacterio)phage cycles. However, in both cases the virus/phage replicates using the host DNA machinery.

The lytic cycle, which is also commonly referred to as the "reproductive cycle" of the bacteriaphage, is a six-stage cycle. The six stages are: attachment, penetration, transcription, biosynthesis, maturation, and lysis.

- Attachment: The phage attaches itself to the surface of the host cell in order to inject its DNA into the cell.
- Penetration: The phage injects its DNA into the host cell by penetrating through the cell membrane.
- Transcription: The host cell's DNA is degraded and the cell's metabolism is directed to initiate phage biosynthesis.
- Biosynthesis: The phage DNA replicates inside the cell, synthesizing new phage DNA and proteins.

- Maturation: The replicated material assembles into fully formed viral phages (each made up of a head, a tail and tail fibers).
- Lysis: The newly formed phages are released from the infected cell (which is itself destroyed in the process) to seek out new host cells to infect.

Attachment and Penetration

To infect a host cell, the virus must first inject its own nucleic acid into the cell through the plasma membrane and (if present) the cell wall. The virus does so by either attaching to a receptor on the cell's surface or by simple mechanical force. The binding is due to electrostatic interactions and is influenced by pH and the presence of ions. The virus then releases its genetic material (either single- or double-stranded RNA or DNA) into the cell. In some viruses this genetic material is circular and mimics a bacterial plasmid. At this stage the cell becomes infected and can also be targeted by the immune system. It is mostly aided by receptors in the surface of the cell.

Transcription and Biosynthesis

During the transcription and biosynthesis stages, the virus hijacks the cell's replication and translation mechanisms, using them to make more viruses. The virus's nucleic acid uses the host cell's metabolic machinery to make large amounts of viral components.

In the case of DNA viruses, the DNA transcribes itself into messenger RNA (mRNA) molecules that are then used to direct the cell's ribosomes. One of the first polypeptides to be translated destroys the host's DNA. In retroviruses (which inject an RNA strand), a unique enzyme called reverse transcriptase transcribes the viral RNA into DNA, which is then transcribed again into RNA. Once the viral DNA has taken control it induces the host cell's machinery to synthesize viral DNA, protein and starts multiplying.

The biosynthesis is (e.g. T4) regulated in three phases of mRNA production followed by a phase of protein production.

Early Phase

Enzymes modify the host's transcriptional process by RNA polymerase. Amongst other modifications, virus T4 changes the sigma factor of the host by producing an anti-sigma factor so that the host promotors are not recognized any more but now recognize T4 middle proteins. For protein synthesis Shine-Dalgarno subsequence GAGG dominates an early genes translation.

Middle Phase

Virus nucleic acid (DNA or RNA depending on virus type).

Late Phase

Structural proteins including those for the head and the tail.

Maturation and Lysis

About 25 minutes after initial infection, approximately 200 new bacteriophages (virions) are formed. Once enough virions have matured and accumulated, specialized viral proteins are used to dissolve the bacterial cell wall. The cell bursts (i.e. it undergoes lysis) due to high internal osmotic pressure (water pressure) that can no longer be constrained by the cell wall. This releases progeny virions into the surrounding environment, where they can go on to infect other cells and another lytic cycle begins. The phage that causes lysis of the host is called a lytic or virulent phage.

Gene Regulation Biochemistry

There are three classes of genes in the phage genome that regulate whether the lytic or lysogenic cycles will emerge. The first class is the immediate early genes, the second is the delayed early genes and the third is the late genes. The following refers to the well-studied temperate phage lambda of E. coli.

- Immediate early genes: These genes are expressed from promoters recognized by the host RNA polymerase, and include Cro, cII, and N. CII is a transcription factor that stimulates expression of the main lysogenic repressor gene, cI, whereas Cro is a repressor for cI expression. The lysis-lysogeny decision is mainly influenced by the competition between Cro and CII, resulting in the determination of whether or not sufficient CI repressor is made. If so, CI represses the early promoters and the infection is shunted into the lysogenic pathway. N is an anti-termination factor that is needed for the transcription of the delayed early genes.
- Delayed early genes: These include the replication genes O and P and also Q, which encodes the anti-terminator responsible for transcription of all the late genes.
- Late genes: Q-mediated turn-on of late transcription begins about 6-8 min after infection if the lytic pathway is chosen. More than 25 genes are expressed from the single late promoter, resulting in four parallel biosynthetic pathways. Three of the pathways are for production of the three components of the virion: the DNA-filled head, the tail, and the side tail fibers. The virions self-assemble from these components, with the first virion appearing at about 20 min after infection. The fourth pathway is for lysis. In lambda 5 proteins are involved in lysis: the holin and antiholin from gene S, the endolysin from gene R and the spanin proteins from genes Rz and Rz1. In wild-type lambda, lysis occurs at about 50 min, releasing approximately 100 completed virions. The timing of lysis is

determined by the holin and antiholin proteins, with the latter inhibiting the former. In overview, the holin protein accumulates in the cytoplasmic membrane until suddenly forming micron-scale holes, which triggers lysis. The endolysin R is released to the periplasm, where it attacks the peptidoglycan. The spanin proteins Rz and Rz1 accumulate in the cytoplasmic and outer membranes, respectively, and form a complex spaning the periplasm through the meshwork of the peptidoglycan. When the endolysin degrades the peptidoglycan, the spanin complexes are liberated and cause disruption of the outer membrane. Destruction of the peptidoglycan by the endolysin and disruption of the outer membrane by the spanin complex are both required for lysis in lambda infections.

- Lysis inhibition: T4-like phages have two genes, rI and rIII, that inhibit the T4 holin, if the infected cell undergoes super-infection by another T4 (or closely related) virion. Repeated super-infection can cause the T4 infection to continue without lysis for hours, leading to accumulation of virions to levels 10-fold higher than normal.

Lysogenic Cycle

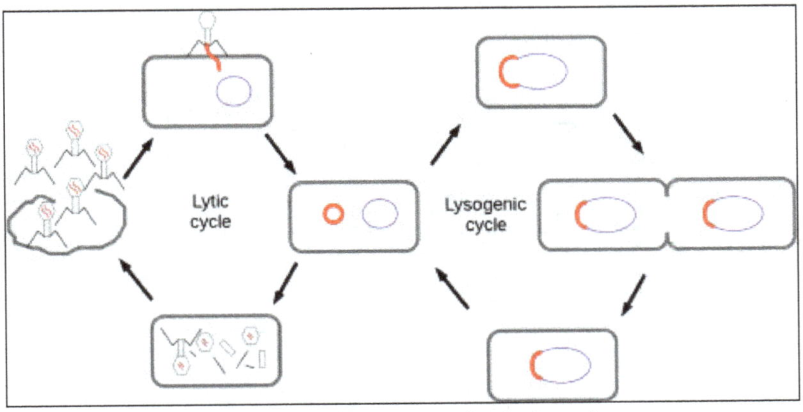

Lysogenic cycle, compared to lytic cycle.

Lysogenic Cycle: 1. The prokaryotic cell is shown with its DNA, in green. 2. The bacteriophage attaches and releases its DNA, shown in red, into the prokaryotic cell. 3. The phage DNA then moves through the cell to the host's DNA. 4. The phage DNA integrates itself into the host cell's DNA, creating prophage. 5. The prophage then remains dormant until the host cell divides. 6. After the host cell has divided, the phage DNA in the daughter cells activate, and the phage DNA begins to express itself. Some of the cells containing the prophage go on to create new phages which will move on to infect other cells.

Lysogeny, or the lysogenic cycle, is one of two cycles of viral reproduction (the lytic cycle being the other). Lysogeny is characterized by integration of the bacteriophage nucleic acid into the host bacterium's genome or formation of a circular replicon in the bacterial cytoplasm. In this condition the bacterium continues to live and reproduce normally. The genetic material of the bacteriophage, called a prophage, can be transmitted to daughter cells at each subsequent cell division, and at later events (such as UV radiation or the presence of certain chemicals) can release it, causing proliferation of new phages via the lytic cycle. Lysogenic cycles can also occur in eukaryotes, although the method of DNA incorporation is not fully understood.

The difference between lysogenic and lytic cycles is that, in lysogenic cycles, the spread of the viral DNA occurs through the usual prokaryotic reproduction, whereas a lytic cycle is more immediate in that it results in many copies of the virus being created very quickly and the cell is destroyed. One key difference between the lytic cycle and the lysogenic cycle is that the lysogenic cycle does not lyse the host cell straight away. Phages that replicate only via the lytic cycle are known as virulent phages while phages that replicate using both lytic and lysogenic cycles are known as temperate phages.

In the lysogenic cycle, the phage DNA first integrates into the bacterial chromosome to produce the prophage. When the bacterium reproduces, the prophage is also copied and is present in each of the daughter cells. The daughter cells can continue to replicate with the prophage present or the prophage can exit the bacterial chromosome to initiate the lytic cycle. In lysogenic cycle the host DNA is not hydrolysed but in lytic cycle the host DNA is hydrolysed in the lytic phase.

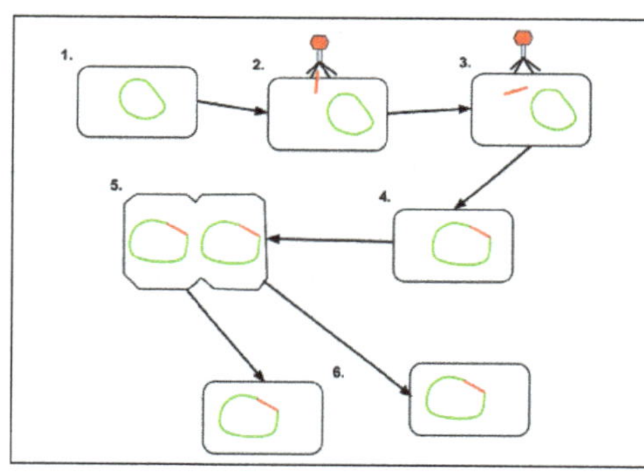

Bacteriophages

Bacteriophages are viruses that infect and replicate within a bacterium. Temperate phages (such as lambda phage) can reproduce using both the lytic and the lysogenic cycle. Via the lysogenic cycle, the bacteriophage's genome is not expressed and is instead integrated into the bacteria's genome to form the prophage. Since the bacteriophage's genetic information is incorporated into the bacteria's genetic information as a prophage, the bacteriophage replicates passively as the bacterium divides to form daughter bacteria cells. In this scenario, the daughter bacteria cells contain prophage and are known as lysogens. Lysogens can remain in the lysogenic cycle for many generations but can switch to the lytic cycle at any time via a process known as induction. During induction, prophage DNA is excised from the bacterial genome and is transcribed and translated to make coat proteins for the virus and regulate lytic growth.

The model organism for studying lysogeny is lambda phage.

Fitness Tradeoffs for Bacteria

Bacteriophages are parasitic because they infect their hosts, use bacterial machinery to replicate, and ultimately lyse the bacteria. Temperate phages can lead to both advantages and disadvantages for their hosts via the lysogenic cycle. During the lysogenic cycle, the virus genome is incorporated as prophage and a repressor prevents viral replication. Nonetheless, a temperate phage can escape repression to replicate, produce viral particles, and lyse the bacteria. The temperate phage escaping repression would be a disadvantage for the bacteria. On the other hand, the prophage may transfer genes that enhance host virulence and resistance to the immune system. Also, the repressor produced by the prophage that prevents prophage genes from being expressed confers an immunity for the host bacteria from lytic infection by related viruses.

Another system, named Arbitrrium, has recently been described for bacteriophages infecting several Bacillus species, in which the decision between lysis and lysogeny is transmitted between bacteria by a peptide factor.

Lysogenic Conversion

In some interactions between lysogenic phages and bacteria, lysogenic conversion may occur, which can also be called phage conversion. It is when a temperate phage induces a change in the phenotype of the infected bacteria that is not part of a usual phage cycle. Changes can often involve the external membrane of the cell by making it impervious to other phages or even by increasing the pathogenic capability of the bacteria for a host. In this way, temperate bacteriophages also play a role in the spread of virulence factors, such as exotoxins and exoenzymes, amongst bacteria. This change then stays in the genome of the infected bacteria and is copied and passed down to daughter cells.

Bacterial Survival

Lysogenic conversion has shown to enable biofilm formation in Bacillus anthracis Strains of B. anthracis cured of all phage were unable to form biofilms, which are surface-adhered bacterial communities that enable bacteria to better access nutrients and survive environmental stresses. In addition to biofilm formation in B. anthracis, lysogenic conversion of Bacillus subtilis, Bacillus thuringiensis, and Bacillus cereus has shown an enhanced rate or extent of sporulation. Sporulation produces endospores, which are metabolically dormant forms of the bacteria that are highly resistant to temperature, ionizing radiation, desiccation, antibiotics, and disinfectants.

Bacterial Virulence

Non-virulent bacteria have also been shown to transform into highly virulent pathogens through lysogenic conversion with the virulence factors carried on the lysogenic prophage. Virulence genes carried within prophages as discrete autonomous genetic elements, known as morons, confer an advantage to the bacteria that indirectly benefits the virus through enhanced lysogen survival.

Examples:

- Corynebacterium diphtheriae produces the toxin of diphtheria only when it is infected by the phage β. In this case, the gene that codes for the toxin is carried by the phage, not the bacterium.
- Vibrio cholerae is a non-toxic strain that can become toxic, producing cholera toxin, when it is infected with the phage CTXφ.
- Shigella dysenteriae, which produces dysentery has toxins that fall into two major groups, Stx1 and Stx2, whose genes are considered to be part of the genome of lambdoid prophages.
- Streptococcus pyogenes, produce a pyrogenic exotoxin, obtained by lysogenic conversion, which causes fever and a scarlet-red rash, scarlet fever.
- Certain strains of Clostridium botulinum, which causes botulism, express botulinum toxin from phage-tranduced genes.

Preventing Lysogenic Induction

Strategies to combat certain bacterial infections by blocking prophage induction (the transition from the lysogenic to the lytic cycle) by eliminating in vivo induction agents have been proposed. Reactive oxygen species (ROS), such as hydrogen peroxide, are strong oxidizing agents that can decompose into free radicals and cause DNA damage to bacteria, which leads to prophage induction. One potential strategy to combat prophage induction is through the use of glutathione, a strong antioxidant that can remove free radical intermediates. Another approach could be to cause an overexpression of CI repressor since prophage induction only occurs when the concentration of CI repressor is too low.

Viroplasm

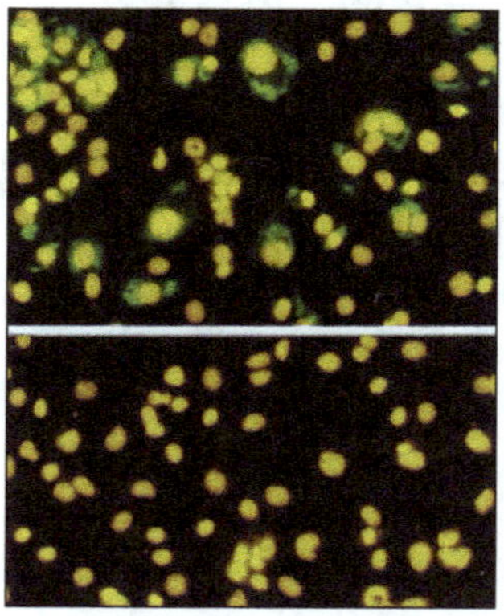

Viroplasms (green) in cells infected with rotavirus (top), and uninfected cells (bottom). (Immunofluorescent stain).

A viroplasm is an inclusion body in a cell where viral replication and assembly occurs. They may be thought of as viral factories in the cell. There are many viroplasms in one infected cell, where they appear dense to electron microscopy. Very little is understood about the mechanism of viroplasm formation.

A viroplasm is a perinuclear or a cytoplasmic large compartment where viral replication and assembly occurs. The viroplasm formation is caused by the interactions between the virus and the infected cell, where viral products and cell elements are confined. Virosomes are also called 'virus factories', and 'virus inclusions'.

Groups of Viruses that Form Viroplasms

Viroplasms have been reported in many unrelated groups of Eukaryotic viruses that replicate in cytoplasm, however, viroplasms from plant viruses have not been as studied as viroplasms from animal viruses. Viroplasms have been found in the cauliflower mosaic virus, rotavirus, vaccinia virus and the rice dwarf virus. These appear electron-dense under an electron microscope and are insoluble.

Baltimore's classification	Family	Species
I: dsDNA viruses	Poxviridae Asfarviridae Iridoviridae Mimiviridae Herpesviridae	Vaccinia virus African swine fever virus frog virus Acanthamoeba polyphaga mimivirus Pithovirus Herpes simplex virus
II: ssDNA viruses		
III: dsRNA viruses	Reoviridae	Avian reovirus
IV: (+)ssRNA viruses	Togaviridae Flaviviridae	Rubella virus Flavivirus
V: (−)ssRNA viruses	Rhabdoviridae	Rabies virus
VI: ssRNA-RT viruses	Retroviridae	Human immunodeficiency virus
VII: dsDNA-RT viruses	Caulimoviridae	Cauliflower mosaic virus

Structure and Formation

Viroplasms are localized in the perinuclear area or in the cytoplasm of infected cells and are formed early in the infection cycle. The number and the size of viroplasms depend on the virus, the virus isolate, hosts species, and the stage of the infection. For example, viroplasms of mimivirus have a similar size to the nucleus of its host, the amoeba Acanthamoeba polyphaga.

A virus can induce changes in composition and organization of host cell cytoskeletal and membrane compartments, depending on the step of the viral replication cycle. This process involves a number of complex interactions and signaling events between viral and host cell factors.

Viroplasms are formed early during the infection; in many cases, the cellular rearrangements caused during virus infection lead to the construction of sophisticated inclusions —viroplasms— in the cell where the factory will be assembled. The viroplasm is where components such as replicase enzymes, virus genetic material, and host proteins required for replication concentrate, and thereby increase the efficiency of replication. At the same time, large amounts of ribosomes, protein-synthesis components, protein folding chaperones, and mitochondria are recruited. Some of the membrane components are used for viral replication while some others will be modified to produce viral envelopes, when the viruses are enveloped. The viral replication, protein synthesis and assembly require a considerable amount of energy, provided by large clusters of mitochondria at the periphery of viroplasms. The virus factory is often enclosed by a membrane derived from the rough endoplasmic reticulum or by cytoskeletal elements.

In animal cells, virus particles are gathered by the microtubule-dependent aggregation of toxic or misfolded protein near the microtubule organizing center (MTOC), so the viroplasms of animal viruses are generally localized near the MTOC. MTOCs are not found in plant cells. Plant viruses induce the rearrangement of membranes structures to form the viroplasm. This is mostly shown for plant RNA viruses.

Functions

Viroplasm is the location within the infected cell where viral replication and assembly take place. Wrapping the viroplasm with a membrane, concentrates the viral components required for the genome replication and the morphogenesis of new virus particles, so it increases the efficiency of the processes. The recruitment of cellular membranes and cytoskeleton to generate virus replication sites can also benefit viruses in other ways. Disruption of cellular membranes can, for example, slow the transport of immunomodulatory proteins to the surface of infected cells and protect against innate and acquired immune responses, and rearrangements to cytoskeleton can facilitate virus release. The viroplasm could also prevent virus degradation by proteases and nucleases.

In the case of the Cauliflower mosaic virus (CaMV), viroplasms improve the virus transmission by the aphid vector. Viroplasms also control release of virions when the insect stings an infected plant cell or a cell near the infected cells.

Possible Co-evolution with the Host

Aggregated structures may protect viral functional complexes from the cellular degradation systems. For example, formation of viral factories of the ASFV viroplasm is very similar to the aggresome formation. An aggresome is a perinuclear site where misfolded proteins are transported and stored by the cell components for their destruction. It has been proposed that the viroplasm could be the product of a co-evolution between the virus and its host. It is possible that a cellular response originally designed to reduce the toxicity of misfolded proteins is exploited by cytoplasmic viruses to improve their replication, the virus capsid synthesis, and assembly. Alternatively, the activation of host defense mechanisms may involve sequestration of virus components in aggregates to prevent their dissemination, followed by their neutralisation. For example, viroplasms of mammalian viruses contain certain elements of the cellular degradation machinery which might enable cellular protective mechanisms against viral components. Given the co-evolution of viruses with their host cells, changes in cell structure induced during infection are likely to involve a combination of the two strategies.

Use in Diagnostics

Presence of viroplasms is used to diagnose certain viral infections. Understanding the phenomena of virus aggregation and of the cell response to the presence of virus, and whether viroplasms facilitate or inhibit viral replication, may help to develop new therapeutic approaches against virus infections in animal and plant cells.

Viral Shedding

Viral shedding refers to the expulsion and release of virus progeny following successful reproduction during a host-cell infection. Once replication has been completed and the host cell is exhausted of all resources in making viral progeny, the viruses may begin to leave the cell by several methods.

The term is used to refer to shedding from a single cell, shedding from one part of the body into another part of the body, and shedding from bodies into the environment where the viruses may infect other bodies.

Budding

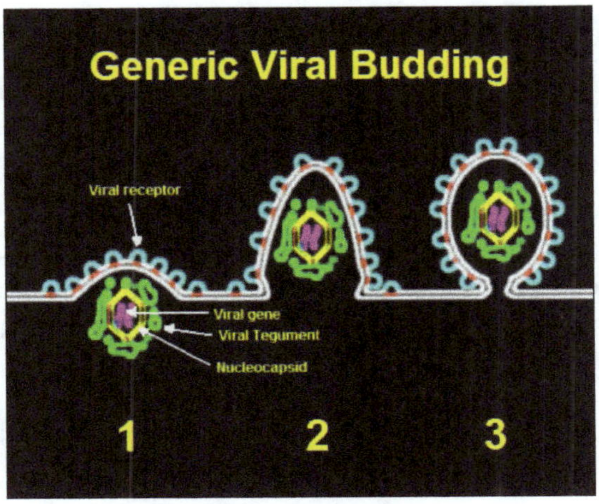

Generic viral budding.

"Budding" through the cell envelope—in effect, borrowing from the cell membrane to create the virus's own viral envelope—is most effective for viruses that need an envelope in the first place. These include enveloped viruses such as HIV, HSV, SARS or smallpox. Prior to budding, the virus may put its own receptor onto the surface of the cell in preparation for the virus to bud through, forming an envelope with the viral receptors already on it. Though budding does not immediately destroy the host cell, this process will slowly use up the cell membrane and eventually lead to the cell's demise. This is also how antiviral responses are able to detect virus-infected cells. Budding has been most extensively studied for viruses of eukaryotes. However, it has been demonstrated that viruses infecting prokaryotes of the domain Archaea also employ this mechanism of virion release.

Apoptosis

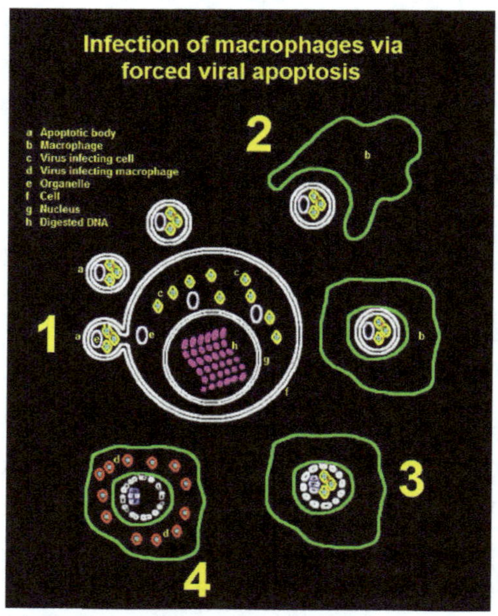

Virus forcing cell to undergo apoptosis to infect macrophages.

Animal cells are programmed to self-destruct when they are under viral attack or damaged in some other way. By forcing the cell to undergo apoptosis or cell suicide, release of progeny into the extracellular space is possible. However, apoptosis does not necessarily result in the cell simply popping open and spilling its contents into the extracellular space. Rather, apoptosis is usually controlled and results in the cell's genome being chopped up, before apoptotic bodies of dead cell material clump off the cell to be absorbed by macrophages. This is a good way for a virus to get into macrophages either to infect them or simply travel to other tissues in the body.

Although this process is primarily used by non-enveloped viruses, enveloped viruses may also use this. HIV is an example of an enveloped virus that exploits this process for the infection of macrophages.

Exocytosis

Viruses also leave cells through exocytosis, in which the host cell is not destroyed. Viral progeny are synthesized within the cell and the host cell's transport system is used to enclose them in vesicles; the vesicles of virus progeny are carried to the cell membrane and then released into the extracellular space. This is used primarily by non-enveloped viruses, although enveloped viruses display this too. An example is the use of recycling viral particle receptors in the enveloped varicella-zoster virus.

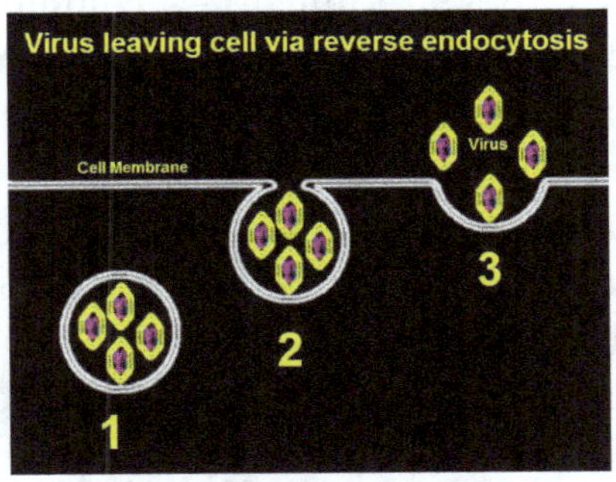

Virus leaving via exocytosis.

Contagiousness

A person with a viral disease is contagious if they are shedding viruses. The rate at which an infected person sheds viruses over time is therefore of considerable interest. Some viruses such as HSV-2 (which produces genital herpes) can cause asymptomatic shedding and therefore spread undetected from person to person as during this kind of shedding, no fever or other hints reveal the contagious nature of the host.

Virus Latency

Viral latency is the ability of a virus to remain dormant within the host cell, sometimes establishing lifelong occult infection. Depending on the virus, the trigger of latency is highly variable but the host cell context is always determining. Latency can stop upon viral genome reactivation, often promoted by stress cellular signals.

Although many viruses are capable of subclinical infection, only a few are known to undergo true latency. In latent infection, the full viral genome is retained in the host cell, but its expression is dramatically restricted, such that few viral antigens and no viral particles are produced. To qualify as latency, this cryptic form of infection must display two additional properties: persistence and reversibility. Reversibility – i.e. the capacity of the genome to, under the appropriate circumstances, reactivate full viral gene expression, with production of infectious progeny (so-called productive or lytic replication) - is the key requirement of latency. Cryptic states that lack this characteristic are more properly characterized as abortive infections, which typically occur when viruses infect cells that are nonpermissive for viral replication. Nonpermissivity can occur due to events at the cell surface – e.g. the absence of receptors or other entry cofactors – or due to post-entry blocks to subsequent steps in viral growth. When post-entry blocks are present (either due to the absence of a permissive factor or the presence of an inhibitory one), abortive infection results.

Abortive infections are well-described in cell culture, but because they cannot lead to persistence or spread in vivo, they are much less commonly identified in the intact host and likely have little biological significance - in essence, they are a dead end. It is the property of reversibility that allows latent infections to avoid a similar fate, and instead become an effective mechanism of viral persistence. Efficient establishment of latency allows the viral genome to persist despite host immune responses to many viral antigens, and in the face of other potentially adverse signals in the microenvironment. When environmental conditions warrant, appropriate signals can trigger the full repertoire of viral gene expression, allowing virus production and spread to resume.

Only a few virus families are known to be capable of true latency, as strictly defined above. Chief among these are the herpesviruses, a huge and widely distributed family of DNA viruses that are important pathogens in their native vertebrate hosts. The capacity for latency is a defining feature of herpesviruses: all known herpesviral infections display latency in every infected individual. Indeed, the anatomic sites of latency, and the frequency with which latency is reversed to engender lytic infection, are important determinants of the clinical manifestations of infection. The other important virus family in which latency has been described are the retroviruses. These small, enveloped RNA viruses replicate via reverse transcription of their RNA genomes; the resulting DNA establishes persistence by integrating into the host genome, from which it generally cannot be dislodged. In most infected cells in culture, such integrants continue to express genomic RNA and viral proteins, leading to the production of progeny virions. However, in some cultured cells such integrants are completely transcriptionally silent, though they can be exogenously stimulated to express viral mRNAs and resume virus production. These cells meet the molecular definition of latency.

In the intact host, the best-characterized example of retroviral latency occurs in HIV infection of humans, in which a small subpopulation of long-lived memory T cells can undergo this form of latent infection. This latency has enormous clinical importance, for "archived" copies of proviral DNA in this locale cannot be eliminated with conventional antiretroviral drugs, and can lead to relapse of infection when such drugs are discontinued. But it is open to question whether such latency is critical for HIV persistence in a natural infection – active viral replication, which is continuously present in many more cells than are latently infected, appears more than adequate to assure this outcome in the host. Human HTLV infection is another case in which true latency likely exists in vivo. In that case, many circulating cells bear proviral DNA in the absence of overt viral replication or gene expression, though the exact role of these cells in the natural history of infection and persistence is still being debated. In most other retroviral infections, the issue of latency in vivo has been little studied. So even though latency clearly exists in retroviral infection, its universality and biological importance in vivo is not as well established for these agents as it is for herpesviruses.

Herpesviral Infection: A Primer

Herpesviruses are large, enveloped DNA viruses that can engender either a latent or a lytic infection at the level of the single cell. Three major subfamilies of herpesviruses – termed α, β and γ - are recognized based on sequence phylogeny. The alphaviruses establish latent infections in neurons, while the gammaherpesviruses are markedly lymphotropic (the tropism of the betaherpesviruses is more variable). Among the alpha- and beta-viruses, lytic infection is the default pathway in culture. Moreover, for none of these viruses has an efficient cell culture model of latency been developed that faithfully recapitulates the cardinal features of latent infection in vivo. As a result, our understanding of lytic replication is very advanced in these viruses, but much less is known about the molecular basis of their latent states. For most of these viruses, latency can be studied only in an intact human or animal host. This poses a formidable experimental barrier to mechanistic understanding of latency, since (i) relatively few latently infected cells exist in any given tissue, and (ii) the extent and level of viral gene expression in latency is generally low.

By contrast, among the gammaherpesviruses, latency is generally (though not invariably) the default pathway in cell culture. As a result, it is usually straightforward to obtain substantial populations of latently infected cells following in vitro infection, facilitating the analysis of both viral and host gene expression in such populations, as well as analyses of the details of viral genome and chromatin structure. In some cases, stably latent cell populations can be continuously cultured, allowing analysis of viral genes and cis-acting elements required for genomic persistence. For all of these reasons, more is known about latency among the lymphotropic subgroup than about latency in any other subfamily of herpesviruses. Accordingly, in this review, we focus on how latency is established and maintained in the gammaherpesviruses, and how these processes are regulated.

Common Features of the Gammaherpesviral Replication Cycle

All gammaherpes viruses encapsidate duplex linear DNA genomes in their particles. Upon infection of cells, this DNA is delivered to the nucleus, where it is circularized (largely by host enzymatic machinery) generating a closed-circular DNA form that can persist in the nucleus as a plasmid. Incoming virion DNA generally lacks histones, but the resulting plasmid is rapidly chromatinized in the nucleus. Since latency is generally the default pathway in this group, most viral gene expression from the plasmid is silenced, but a handful of genes are expressed from the latent genome. As a general rule, one of more of these genes directs replication of the viral genome, using a cis-acting sequence (termed oriP) as the replication origin of the plasmid. Several latency genes are involved in modulation of host signaling, most notably influencing the activation of NFκB, which turns out to be important in the maintenance of latency. Interestingly, however, these common functions are often carried out by viral proteins that share little homology across the different gammaherpesviruses. For example, none of the latency proteins of EBV are conserved in KSHV or MHV68 (though the latter two viruses do share a number of latency genes).

Latency is reversible. Because the full complement of viral DNA is retained in the nucleus, under the appropriate circumstances the second program of viral gene expression, lytic replication, can be activated. In this program, expression of virtually all viral ORFs is activated, in a temporally regulated cascade. Three major waves of viral gene expression occur: immediate early (IE) genes, the first wave, are predominantly regulators of gene expression; they turn on delayed early (DE) genes, many of which are functions involved in DNA replication, signal transduction, shutoff of host macromolecular synthesis and immune evasion. DE expression triggers robust viral DNA synthesis - which is now delinked from that of the host, and proceeds by a rolling-circle mechanism. Following DNA replication, late (L) genes are activated; these are mostly structural components of the virion, and their accumulation leads to encapsidation of newly replicated viral genomes into progeny virions. In contrast to the latency genes, there is extensive conservation of lytic cycle genes across all herpesviruses. In all three viruses under review here, lytic reactivation is controlled by one or more (usually one) master regulators of transcription. Expression of these so-called lytic switch protein genes is silenced in latency, but is turned on by all signals that trigger lytic reactivation; this activator, in turn, triggers expression downstream genes to kick-start the lytic cycle.

Gammaherpesviruses thus face a number of common problems that must be solved in order for latency to be established and maintained. First, incoming viral genomes must be chromatinized, and epigenetic and genetic controls established that allow for the latent transcriptional program. Second, latency functions must provide for stable replication from oriP, and for mechanisms to ensure segregation of the viral genome to daughter cells if the latently infected cell should divide. Third, latent regulators of signaling must create an environment that fosters the stability of latency without rendering it irreversible. Part of this involves establishing genetic and epigenetic controls over the expression of the lytic switch protein, as well as biochemical controls over its function. Finally, in the intact host successful latency may also require modulation of host cell functions, particularly those affecting the lifespan and proliferative potential of the latently infected cell. Our molecular understanding of all of these issues is still very incomplete, but some unifying themes are starting to come into view. Even though the latency proteins are poorly conserved among the viruses, commonalities in the modes by which they promote episome maintenance and in the signaling pathways they engage are emerging. In what follows, we discuss in detail the latency strategy of three individual gammaherpesviruses, with particular attention to these themes - and their variations.

EBV Latency and its Control

EBV was the first identified gammaherpesvirus and remains the best characterized member of this family. Initially recovered from cell lines derived from African Burkitt's lymphoma (BL) in the 1960s, EBV has since been associated with a variety of lymphoproliferative diseases, including infectious mononucleosis, 30-40% of Hodgkin's disease, and the majority of cases of post-transplant lymphoproliferative disease (PTLD). In addition, EBV infection is also strongly linked several nonlymphoid disorders, including nasopharyngeal carcinoma and AIDS-related oral hairy leukoplakia. In vitro and in vivo, the primary target of EBV infection is the B cell; following de novo infection of B cells in culture, latency is the default pathway. In vitro infection of primary B cells often results in immortalization, and the resulting lymphoblastoid cell lines (LCLs) have been an important source of information about EBV latency. EBV latency is a complex subject that can present several different faces, not all of which are captured by the study of LCLs. Latently infected cells can also be derived from EBV-associated tumors, which often display different patterns of latent gene expression.

In EBV, the first regulator of the latent-lytic switch to be discovered was the protein Zta (aka BZLF1), a bZIP transcription factor that is centrally involved in turning on the subsequent genes of the lytic cycle. Subsequently, a second viral protein, Rta, was also found to be essential for induction of the lytic cascade. Both Zta and Rta are transcriptional activators; they can each upregulate the other's expression, and together act to turn on downstream lytic genes. Ectopic expression of either gene can induce latently infected cells to commence lytic EBV production, and inactivation of either gene blocks lytic reactivation. Rta is well conserved among all the known gammaherpesviruses, while Zta is less well conserved. Typically, lytic replication is induced by chemicals like phorbol esters or sodium butyrate, which are thought to act by activating Zta expression. Thus, regulation of Zta expression and function is a key determinant of the latent-lytic switch.

Upon de novo infection, incoming virion DNA is largely unmethylated. This is important because Zta binds preferentially to methylated sites at many of its recognition elements; as such, even though there is a burst of Zta synthesis following de novo infection of B cells, this does not trigger immediate entry into the lytic cycle. As the EBV genomes becomes chromatinized, CpG methylation proceeds; however, by the time most Zta target sites are methylated, expression of Zta has been quelled by the formation of repressive complexes on the BZLF1 promoter - most notably through binding of phosphorylated MEF2D (which binds to 3 sites within the Zta promoter) and recruitment of histone deacetylases. Thus, the epigenetic marking of the EBV genome plays a central role in regulating virus replication during the establishment of latency. It should also be pointed out that the core promoter driving Zta expression is devoid of CpG dinucleotides and thus is never methylated. This may be a key feature that allows the initial expression of Zta following appropriate reactivation stimuli.

A number of factors can inhibit the latent-lytic switch and thereby stabilize latency. Chief among these is the activation of the transcription factor NFkB. The p65 subunit of this transcription factor can interfere with Zta activation of target promoters; this inhibition is mediated through a direction interaction of p65 with Zta. Similar antagonism of lytic reactivation by active NFkB is also seen in KSHV and MHV68 infection.

What are the in vivo cues that signal EBV to reactivate from a latently infected B cell? Data from several groups has now shown that stimulation of plasma cell differentiation leads to virus reactivation. The critical cellular factor involved in initiating EBV reactivation is XBP-1s, which normally plays a central role in the cellular unfolded protein response. XBP-1s expression is induced by BLIMP-1, the master regulator of plasma cell differentiation. XBP-1s is a member of the CREB family of transcription factors and binds to a site in the EBV Zta promoter that has previously been shown to bind CREB and AP-1 proteins. Thus, EBV is apparently able to usurp plasma cell differentiation to turn an antibody producing cell into a virus factory.

The big remaining question is: what regulates differentiation of latently infected B cells to plasma cells? At present, the answer to this is unknown. Although cognate antigen recognition is one such stimulus, in the case of MHV68 a viral antigen has been identified that can drive plasma cell differentiation. This suggests the tantalizing possibility that plasma cell differentiation of latently infected B cells may be regulated by the virus rather than by antigen stimulation. But many other possibilities remain, and it is conceivable that both viral and host factors modulate this decision.

EBV Latency Program

Growth Program

Shortly after the identification of EBV in Burkitt's lymphoma (BL) tumors, it was shown that the virus could efficiently immortalize primary B cells to LCLs. This discovery had a huge impact on the field, since it (i) provided a ready source of latently infected B cells for study; and (ii) appeared to cement the relationship between EBV infection and the genesis of BL. The next 20 years were largely spent identifying and characterizing viral genes expressed in such LCLs. This work led to the identification of 9 viral proteins, along with several non-coding RNAs, that were consistently detected. Because these gene products trigger B cell proliferation, they are sometimes referred to as the EBV "growth program." The viral proteins expressed can be broken into 2 distinct groups - 6 nuclear antigens (Epstein-Barr nuclear antigens, EBNAs) and 3 membrane antigens (latency-associated membrane proteins, LMPs). The viral genes expressed during latency are not clustered in a single region, but rather are spread out over most of the viral genome. However, the promoters and cis-elements controlling these genes are in fact clustered in a relatively small region spanning the fused terminal repeats of the viral episome.

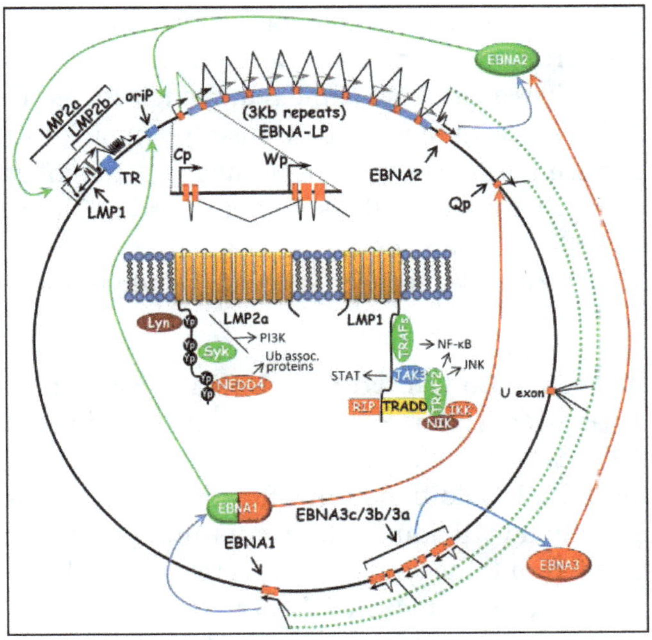

Schematic illustration of EBV EBNA and LMP gene transcription, and auto-regulation of viral latency-associated gene expression by the EBNA gene products. An exploded view of Cp- and Wp –driven EBNA gene transcription is shown, depicting the organization of exons immediately downstream of each promoter. Cp-initiated transcript contain 2 unique exons, C1 and C2, which splice to a variable number of W1 and W2 exons encoded within the 3.0Kb internal repeats. Wp-initiated transcript contain a single unique exon, W0, which splices to a variable number of W1 and W2 repeat exons. Also shown in the inset is the predicted membrane topology of the LMP-1 and LMP-2a proteins, along with known cellular interacting partners and signaling pathways activated by these proteins.

An early indication of the complexity of EBV latency-associated gene expression came from the characterization of EBNA gene transcription. These studies revealed that the EBNA gene transcripts all share common 5′ exons that are alternatively spliced to coding exons located at the 3′ end of these transcripts. The exception is EBNA-LP, which is encoded by the common 5′ exons shared among all EBNA gene transcripts. The presence of the EBNA-LP translation initiation codon is determined by alternative splicing and thus both EBNA-LP coding and EBNA_LP- non-coding transcripts are generated - presumably ensuring efficient translation of downstream EBNA gene coding sequences present in the latter RNAs.

During the initial phase of EBV infection, EBNA gene transcription is initiated from a promoter, Wp, located within the major internal repeat of the virus (IR1; 3Kb repeat length). There are multiple copies of this promoter and these collectively appear to function as an "ignition switch" to launch EBNA gene transcription in the newly infected resting B cell. Expression of the most proximal EBNA genes, EBNA-LP and EBNA2, are the first viral proteins expressed and function together to drive a switch in promoter utilization to a second promoter, termed Cp, located just upstream of the IR1 repeat region. This promoter drives expression of the remaining EBNAs. This switch is driven by the interaction of EBNA2 with the cellular transcription factor RBP-Jk (also known as CBP-1 and CSL), which normally functions in the regulation of genes in the cellular Notch signaling pathway. Two other EBNAs, EBNA3A and EBNA3C, also interact with RBP-Jk and displace EBNA2 – thus serving as modulators of EBNA2's transcriptional regulatory activity. EBNA2 interaction with RBP-Jk also serves to turn on transcription of the LMP genes.

An important product of the Cp transcripts is EBNA-1, which plays a central role in latency by binding to sequences in ori P and promoting initiation of DNA replication of the viral episome, which is largely effected by cellular enzymes and proceeds in concert with host DNA replication. In addition to this function, EBNA-1 also plays important roles in plasmid segregation in dividing cells, via its ability to affiliate with metaphase host chromosomes. As a result, the viral genome is tethered to host chromosomes and can therefore be passively distributed with them to daughter cells by the workings of the mitotic spindle. Mutational lesions in regions of EBNA-1 that are responsible for this tethering cannot establish stable episomes in dividing cells.

In addition to its replicative functions, EBNA-1 is also a transcriptional activator. When bound to oriP sequences, which are just upstream of Cp, it can upregulate transcription from this promoter. Thus, the switch from Wp to Cp reflects a shift from utilizing a viral promoter regulated by cellular transcription factors to one which is tightly controlled by viral factors.

Which of the 9 latency-associated proteins are required for EBV immortalization of primary human B cells? Multiple studies have shown a requirement for EBNA1, EBNA2, EBNA3a, EBNA3c and LMP-1, while EBNA3b and LMP2a/b appear dispensable. No attempts to generate an EBNA-LP null virus have been reported, due to the difficulty of knocking out all possible alternatively spliced products that can/could encode a functional form of EBNA-LP. However, deletion of the unique C-terminal 45aa of EBNA-LP has been shown to significantly impair LCL formation.

With respect to transforming activity, only one of the EBV latency-associated antigen, LMP-1, has been directly demonstrated to be an oncogene, as judged by standard growth transformation assays on fibroblasts. LMP-1 is a member of the TNF-receptor superfamily, and appears to function as a constitutively active CD40 receptor. LMP-1 is a ligandless receptor that constitutively trimerizes and recruits a number of cytoplasmic factors [e.g., TNF receptor associated factors (TRAFs)] involved in mediating receptor signaling. Most notably, LMP-1 expression in B lymphocytes strongly upregulates NF-κB activity. A LMP-1 null EBV is unable to immortalize primary human B cells, although it has been shown that LMP-1 function can be replaced by continued growth of LMP-1 null EBV infected B cells on fibroblasts feeder layers.

More recent studies have shown that the episomal maintenance protein EBNA1 is also dispensable for B cell immortalization, although EBNA1 null EBV was several thousand fold less efficient in generating LCLs. As expected, based on the known function of EBNA1 in maintaining the viral episome, all of the LCLs generated with EBNA1 null virus contained integrated EBV genomes. Integration of the viral genome would eliminate the requirement for EBNA1 in maintenance of the viral episome, but not its proposed role in enhancing LMP and EBNA gene transcription. Thus, it seems likely that in LCLs generated with EBNA1 null EBV there would be selection for integration events that compensate for the role of EBNA1 transcriptional activation of EBV latency-associated gene expression. The rarity of such events likely accounts for the sharp reduction in the frequency of LCL outgrowth with EBNA-1 null mutants.

Alternative Latency Programs

While the early studies on EBV immortalized lymphoblastoid cell lines and established BL tumor cell lines supported the notion of a direct role for EBV-driven B cell proliferation in the genesis of BL, a more careful analysis of fresh BL biopsies and early passage BL cell lines revealed a much more restricted pattern of EBV gene expression. In the vast majority of BL tumors the only viral antigen that can be detected is EBNA1 (this is now referred to as type I latency). Subsequent analyses of both nasopharyngeal carcinomas and EBV-positive Hodgkin's lymphomas revealed another restricted pattern of viral gene expression, in which LMP-1 and LMP-2 expression along with EBNA1 could be detected (now termed type II latency; the larger, growth-promoting program of LCLs is now designated type III latency). In all of these tumors, the common theme was expression of EBNA1 in the absence of the other EBNA genes. Since all EBNA gene transcription in EBV LCLs is driven from a single transcriptional unit, this raised the question of how the EBNA1 only programs were regulated. This ultimately led to the identification of an alternative promoter, Qp, located downstream of the EBNA2 coding exons. Notably, Qp-initiated transcripts splice exclusively to the EBNA1 coding exon - bypassing the exons encoding the EBNA3 family of nuclear antigens. How this selectivity in splice site selection is achieved remains a mystery. While EBNA1 upregulates EBNA and LMP gene expression through binding to oriP, it potently inhibits transcription from Qp. A genome wide survey for EBNA1 binding sites in the EBV genome revealed only two sites outside oriP - both of which map just downstream of the Qp transcription initiation site. Unlike Cp and Wp, Qp is a TATA-less promoter that appears related to promoters found upstream of housekeeping genes - suggesting that it may function as the default promoter to ensure ongoing expression of EBNA1 in the absence of transcription initiation from Cp and Wp.

Given what we know about autoregulation of EBNA gene expression, how is Cp/Wp-initiated EBNA gene transcription silenced in the EBV-associated tumors that arise in immunocompetent individuals? A large number of studies have shown that Cp and

Wp are heavily methylated in these tumors, while Qp remains unmethylated and active. That DNA methylation might be involved in modulating viral gene expression was not particularly surprising, since it had been shown that the EBV genome has undergone extensive CpG suppression during its evolution - presumably as a consequence of methylation of the viral genome. Furthermore, the importance of DNA methylation in the suppression of Cp/Wp transcriptional activity in BL cell lines is supported by three independent experimental observations: (i) the drift of some BL cell lines to expression of all the EBNA and LMP genes (type III latency) correlated with the loss of methylation around Cp and Wp; (ii) inhibition of DNA methyltransferase activity induces transcription from Cp/Wp in BL cell lines exhibiting the type I latency program; and (iii) reporter constructs containing extensive sequence upstream and downstream of Cp are active when transfected into BL cell lines exhibiting the type I latency program. The latter result argues that cells in latency type I contain the necessary transcription factors to drive Cp/Wp-initiated transcription, but are prevented from doing so because the viral genomes are methylated and in a transcriptionally inactive conformation.

Since the EBV growth/immortalization program isn't used in EBV-associated tumors, this raises the question of whether EBV is really causally associated with tumorigenesis in these cases – a concern hightened by the discovery that forms of BL exist that are unassociated with EBV infection. In fact, the common feature shared by all cases of BL is not the presence of EBV DNA but the translocation of the c-myc gene to one of the immunoglobulin loci, leading to dysregulation of c-myc expression. Although compelling data exist to link EBV infection with the development of BL, HD and NPC, it is now clear that the role for EBV in the generation of EBV-associated tumors is more complex than originally appreciated.

EBV Latency in its Natural Context

All of the above studies reflect analyses of EBV latency in the highly abnormal context of EBV-induced tumors or LCLs. But what does latency look like in the majority case - i.e. persistent EBV infection of healthy seropositive individuals? Thorley-Lawson and colleagues have shown that in the peripheral blood of such individuals, EBV is found exclusively in resting memory B cells (CD20+, CD27+, CD5-, CD10-, IgD-), and not in activated lymphoblasts. Analysis of viral gene expression in these resting memory B cells failed to detect expression of the viral antigens associated with the EBV growth program (type III latency), save for the sporadic detection of EBNA1 transcripts which are initiated from Qp. The latter appear to correspond to circulating memory B cells that have recently undergone a round of cell division. Whether these EBV infected memory B cells are antigen experienced remains unresolved - however, it has been determined that they have undergone both isotype switching and somatic hypermutation, suggesting antigen selection.

How does EBV get into the memory B cell pool? Examination of tonsil tissue has revealed the presence of EBV infection in: (i) naive B cells (CD19+, sIgD+) which exhibit a lymphoblastoid phenotype and display type III latency; (ii) germinal center B cells which exhibit a type II latency program; (iii) proliferating memory B cells (EBNA1 only; type I latency); and (iv) plasma cells, in which EBV enters the lytic cycle, as discussed earlier. Thus, EBV infection in germinal center B cells largely recapitulates the viral gene expression program in EBV-positive Hodgkin's lymphoma tumors, while EBV infection in cycling memory B cells recapitulates the latency program observed in endemic BL tumors. Based on these snap shots of EBV gene expression in distinct B cell populations, it has been proposed that EBV hijacks normal B cell differentiation to gain access to the memory B cell pool. This model postulates that virus infection of naive B cells results in the generation of activated lymphoblasts that appear phenotypically similar to antigen activated lymphoblasts. Some of these virus infected lymphoblasts then form and participate in germinal center reactions, in which LMP-1 expression provides signals that mimic CD4+ T cell help (CD40 signaling) while LMP-2a provides signals that mimic the essential survival signals normally provided by the B cell receptor - allowing EBV infected B cells to transit through the germinal center reaction. The only other viral antigen expressed in EBV infected germinal center B cells is EBNA1, which serves to maintain the viral episome in proliferating germinal center B cells. From these germinal centers, EBV infected memory B cells and plasma cells emerge. The memory B cells that exit the lymphoid tissue and enter the periphery then stop cycling and shut off EBNA1 gene transcription - presumably due to the accumulation of EBNA1 which negatively feedbacks on Qp, shutting down EBNA1 gene transcription. Analysis of viral genome methylation in B cells isolated from peripheral blood of healthy seropositive individuals revealed that, like the status of the viral genome in BL and Hodgkin's lymphoma tumors, Cp and Wp are methylated while Qp is protected from methylation.

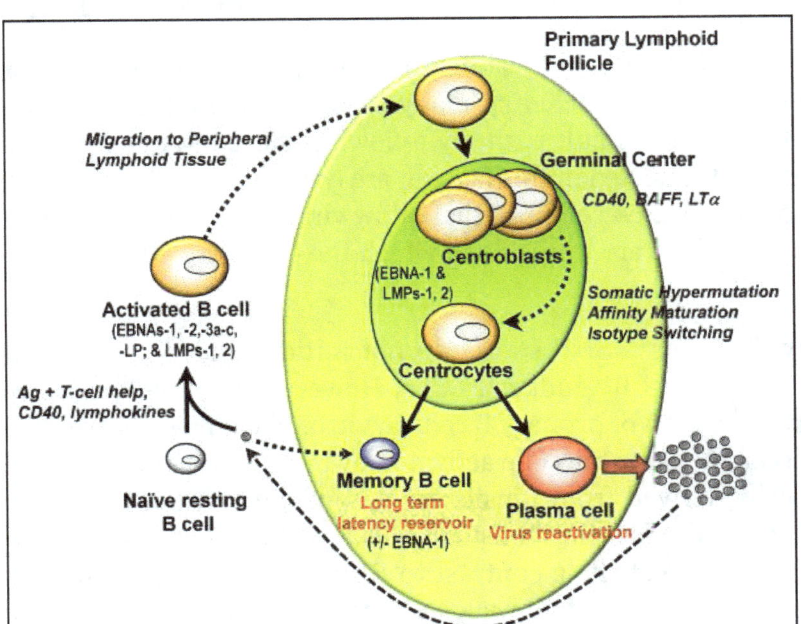

Model depicting the different EBV latency programs expressed during the progression of infected naïve B cells through a germinal center reaction and establishment of latency in memory B cells. Also shown is virus reactivation linked to plasma cell differentiation.

EBV has evolved a complex set of latency programs that allows the virus to navigate from infection of a naïve resting B cells (the dominant B cell population that the virus encounters during the initial stages of infection) to ultimately gain access to the memory B cell reservoir. While there might be a role for antigen in this process, the simplest model involves EBV driving the differentiation of B cells in the absence of any normal signals from the B cell receptor or CD4+ T cells. In this process, EBV takes advantage of an ancient host antimicrobial strategy – host cell-driven DNA methylation – to strategically silence viral gene expression as infection progresses from a naïve B cell to the germinal center reaction. In addition to methylation of the viral genome, modulation of NF-κB activity and manipulation of the Notch signaling pathway through interaction with the host DNA binding protein RBP-Jκ/CBF-1/CSL, represent two key pathways targeted by EBV. The end goal, establishing latency in memory B cells, clearly affords the virus a long lived cell type in which to persist and avoid immune detection.

KSHV Latency and its Control

KSHV (also called human herpesvirus 8) was discovered in 1994 in a search for viral agents in Kaposi's sarcoma, a neoplasm of endothelial cells. Subsequent epidemiologic research affirmed the etiologic link to KS, but also revealed that phylogenetically, KSHV belongs to the lymphotropic herpesviruses. This triggered a search for KSHV DNA in a variety of lymphoproliferative diseases, and led to its linkage to two disorders of B cells: primary effusion lymphoma (PEL) and multicentric Castlemen's disease (MCD). Although rare sequelae of KSHV infection, they reflect the fact that in healthy seropositive hosts, CD19-positive B cells appear to be the primary target of KSHV infection.

KSHV can efficiently infect many adherent cell types in culture (endothelial cells, fibroblasts, epithelial cells), including many cells in which infection is never observed in vivo. As in EBV, the default pathway following these in vitro infections is latency. Although some cells infected in this fashion are lytically inducible, many are not - indicating that not all cultured cell lines that allow viral entry support true latency. Nonetheless, numerous lines are available in which the complete viral replicative cycle can be observed.

Paradoxically, however, KSHV stocks do not initiate infection of established B cell lines – e.g. BL lines or EBV-induced LCLs. However, recently two groups have been able to observe infection of primary B cells in vitro. If such B cells are derived from the peripheral blood, they must first be activated by treatment with IL4 and CD40 ligand. However, B cells derived from tonsils can be infected without such pretreatments, probably reflecting the fact that tonsillar cells are already highly activated in vivo. In both situations, and in striking contrast to EBV infection of primary B cells in vitro, no immortalization of B cells follows KSHV infection. In fact, KSHV latency in all cell types lacks immortalizing activity – indicating that KSHV has no analog of the EBV growth program (latency III). Indeed, none of the known EBV latency genes have homologs in KSHV.

The KSHV genome, as extracted from the virion, is a linear duplex of 165kb. Coding regions bearing at least 87 open reading frames (ORFs) comprise the central 140kb of the genome, and are flanked by extensive, noncoding, GC-rich repeats (terminal repeats, TRs). Following infection, as in EBV, the genome circularizes in the nucleus and, in latently infected cells, is maintained as a chromatinized nuclear plasmid. Latent DNA replication proceeds from an origin (ori P) in the TRs, and the genome is maintained at relatively low copy numbers.

In KSHV, a single viral gene, (RTA, for replication and transcription activator) controls the switch from latency to lytic growth. RTA is a sequence-specific DNA binding protein that can function as a transcriptional activator, and null mutations engineered into full-length KSHV genomes result in viruses that cannot reactivate from latency. RTA's transactivation function controls reactivation by turning on numerous IE and DE promoters, in a fashion formally analogous to the role of Zta in EBV. (KSHV does encode a distant homolog of Zta, but this protein, termed RAP or K8, functions primarily in lytic-cycle DNA replication in KSHV). High-affinity RTA binding sites exist in the promoters of several key lytic-cycle genes, and also near the origins of lytic-cycle DNA replication (where it is presumed that RTA binding promotes an open chromatin configuation favorable to replication). However, many genes that are activated by RTA lack such sites. It is thought that for most such sites, RTA is recruited to the promoter via protein-protein interactions with other, predominantly host-encoded, transcription factors. Chief among these is RBP-Jκ or CSL-1, the main effector of the Notch signaling pathway; ablation of RBP-Jκ does not appear to affect establishment of latency, but efficiently blocks lytic reactivation. This is a very different role for RBP-Jk than is observed in EBV latency; however, at least one report suggests that a latent KSHV protein may also interact with RBP-Jk. RTA also interacts with C/EBP, Oct-1, KRBP and other transcription factors to target the protein to additional promoters, as well as to components of the pol II-associated transcription machinery and to histone acetyl transferases.

Many environmental signals can trigger the switch from KSHV latency to lytic growth in vitro – phorbol esters, histone deacetylase inhibitors, interferon gamma and other cytokines, proteasome inhibitors, inhibitors of NFkB activation, upregulation of protein kinase A or pim-1 and pim-3 kinases, expression of trancription factor XBP-1s; dopaminergic agonists and many others. It is striking that this list so extensively reduplicates the list of signals that trigger EBV reactivation – TPA, HDAc inhibitors, NFkB inhibition and XBP-1s induction all disrupt latency in both pathogens. Although the physiologic signals that trigger lytic KSHV reactivation in vivo are unknown, we do know that periodic "spontaneous" reactivation from latency occurs regularly, both in cell culture and in vivo.

Presumably, all of the above inducers of lytic reactivation act by initiating biochemical events that ultimately converge on activation of the RTA promoter. (Don – Zta promoter lacks CpG and thus methylation doesn't appear to play a role - but EBV Rta promoter is methylated, so there is a parallel there), this promoter is methylated during latent infection, but undergoes demethylation in response to signaling events linked to induction. It seems likely that other epigenetic marks (e.g. histone modifications) are also involved in regulation of RTA expression, but this is an area that has been relatively understudied. It may also be that reactivation is subject to additional controls beyond induction of RTA transcription. For example, RTA is heavily phosphorylated, and it is certainly possible that modulation of this or other post-translational modifications of RTA play adjunctive roles in regulating the efficiency with which latency is disrupted. RTA is also subject to regulation by viral proteins and miRNAs expressed during latency; these will be considered in the following section.

Latency Program of KSHV

Historically, definition of the latent KSHV transcriptional program has largely been carried out by studying B cell lines derived from primary effusion lymphomas (PEL). This led to recognition of a major latency locus that is abundantly and consistently transcribed in all latently infected cells. This region includes four open reading frames, encoding LANA (latency-associated nuclear antigen), v-cyclin, v-FLIP (Flice-inhibitory protein) and kaposins A, B and C. The first three genes are under the control of a single promoter, (the LANA promoter or LTc) which generates a series of coterminal mRNAs via differential splicing. A second promoter (the kaposin promoter or LTd), located just downstream of LANA, encodes a spliced transcript encoding the kaposins, and can also generate a bicistronic RNA for v-cyclin and v-FLIP. This promoter also governs the expression of 12 pre-miRNAs, which can be processed to yield a total of 18 mature miRNAs. All of these latent products have been found to be expressed in KS spindle cells as well as PEL cells.

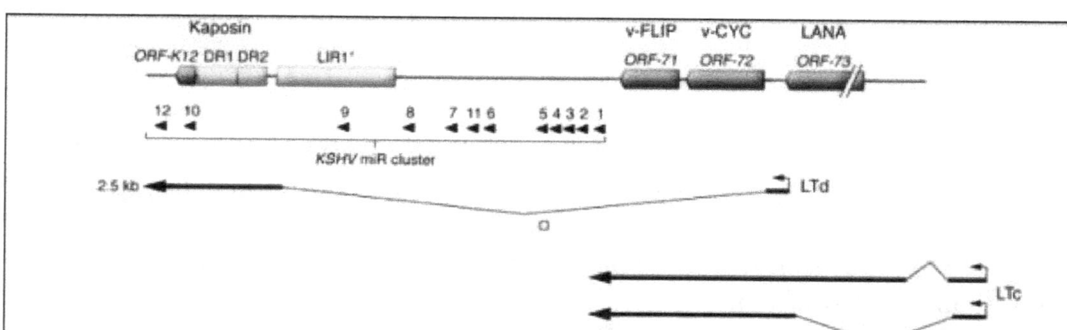

Major latency locus of KSHV. Top line: major ORFS of the locus. ORF-73 encodes LANA; ORF-72 encodes v-cyclin (v-CYC); ORF-71 encodes v-FLIP; ORF-K12 encodes kaposin A; DRs 1 and 2 encode direct repeats in which translation of kaposins B and C initiate. LIR, long interspersed repeats of unknown function. Middle panel: KSHV microRNA (miR) cluster, with pre-miRNAs indicated by arrowheads. Bottom panels: Structures of mRNAs directed by the kaposin (or LTd) promoter and by the LANA (or LTc) promoter.

References

- The-viral-life-cycle, microbiology-chapter: courses.lumenlearning.com, Retrieved 19 July, 2020
- Lakadamyali, Melike; Michael J. Rust; Hazen P. Babcock; Xiaowei Zhuang (2003). "Visualizing infection of individual influenza viruses". Proceedings of the National Academy of Sciences of the United States of America. 100 (16): 9280–9285. doi:10.1073/pnas.0832269100. PMC 170909. PMID 12883000
- Geo. F. Brooks, M.D et al. "Jawetz, Melnick & Adelberg's MEDICAL MICROBIOLOGY.pdf, 26th Edition, McGraw Hill, 2013, ISBN 978-0-07-181578-9
- Malys N (2012). "Shine-Dalgarno sequence of bacteriophage T4: GAGG prevails in early genes". Molecular Biology Reports. 39 (1): 33–9. doi:10.1007/s11033-011-0707-4. PMID 21533668
- Louis-Charles Fortier; et al. (23 April 2013). "Importance of prophages to evolution and virulence of bacterial pathogens". Virulence. 4 (5): 354–65. doi:10.4161/viru.24498. PMC 3714127. PMID 23611873
- Nilsson; Von Bonsdorff, CH; Weclewicz, K; Cohen, J; Svensson, L (1998). "Assembly of viroplasm and virus-like particles of rotavirus by a Semliki Forest virus replicon". Virology. 242 (2): 255–65. doi:10.1006/viro.1997.8987. PMID 9514960
- Pornillos O, Garrus JE, Sundquist WI (December 2002). "Mechanisms of enveloped RNA virus budding". Trends in Cell Biology. 12 (12): 569–79. doi:10.1016/s0962-8924(02)02402-9. PMID 12495845

Viral Proteins

5

- **Capsid**
- **Major Capsid Protein VP1**
- **Minor Capsid Proteins VP2 and VP3**
- **Vpu Protein**
- **Structural Viral Proteins**

Viral proteins are the products of viruses that are grouped as per their functions and structures. It also includes structural proteins, nonstructural proteins, regulatory, and accessory proteins. This chapter delves into the subject of viral proteins for an in-depth understanding of the subject.

Viral proteins are proteins generated by a virus. As viruses hijack much of their host's cellular machinery to support their life cycle, they encode very few of their own genes; viral proteins are therefore generally structural components, for the viral envelope and capsid. Non-structural proteins and regulatory or accessory proteins can also be viral proteins.

Viral proteins can produce epigenetic alterations, and it is hypothesized that these are mainly responsible for the carcinogenesis induced by viruses. DNMTs are key enzymes in inactivating transcription by DNA hypermethylation and ensuring proper gene repression. In order to maintain a harmonious internal environment, the protein level or enzymatic activity of DNMTs should be tightly controlled. Nevertheless, disruption of this delicate balance results in chaos. The fact that EBV LMP1, as well as HBV HBx, HPV E7, and KSHV LANA, activates DNMTs highlights the importance of viral proteins as initiators or co-factors to induce epigenetic alterations. In fact, LMP1 is the first viral gene that was identified as having the ability to activate DNMTs via a defined signaling pathway. Upon SV40 infection, the Ras pathways are activated by T antigen: this in turn activates de novo DNA methyltransferase 3B.

Viral proteins may be either structural or enzymatic. Structural proteins are the "building blocks" used in construction of the protective protein capsule or "capsid" that, in combination, constitute viral particles. Very simple viruses (e.g. the bacteriophage ϕX174, produce as few as two of these structural proteins while complex viruses may encode dozens. Some protein synthesis inhibiting drugs (e.g. puromycin) effectively reduce progress of viral infections. These drugs cause production of defective structural proteins in adequate numbers to inhibit viral assembly. It is notable that some structural proteins appear capable of catalyzing their own assembly into capsids, further illustrating the efficient nature of viral parasites.

Viral proteins can be detected in the early stages by direct assays such as Western blots, though as noted above, much of the protein seen here is not on a productive track. The viral proteins CA, NC, RT, and IN are present in very large, high-molecular weight complexes, and the functional subset is in association with the viral RNA and DNA. The size and structure of these complexes can be evaluated by sedimentation analysis after gentle lysis at various times after infection, with the proteins detected by Western blot. Depending on the virus and conditions of isolation, matrix, RT, and IN can be shown to be associated with viral nucleic acids in these complexes. A substantial portion of the viral CA protein is also found in rapidly sedimenting forms, and as infection proceeds, more and more CA is stripped away into low-molecular weight, slowly sedimenting forms. The size of the RTC thus gradually shifts in its sedimentation properties from as large as 560 S to as small as 100 S. Thus, the process of uncoating is marked by the gradual loss of CA protein from the core, and the timing of that loss is orchestrated to promote efficient infection. Many CA mutants affect the efficiency and timing of core disassembly.

Changes in the state of the virion core provide the conceptual basis for the so-called "fate of capsid" assay, in which the disassembly of the virus is monitored by whether or not the capsid and genomes in a lysate are pelletable through a layer of a sucrose solution. Various CA mutants show distinctive behavior in these assays, revealing the fact that some mutants are slow to uncoat or dissociate from the complex ("hyperstable" mutants), whereas others uncoat prematurely ("unstable" mutants). The hyperstable mutants exhibit a stiffer coat as measured by atomic force microscopy. The correct time of uncoating is crucial for efficient infection, and mutants that are too early or late in uncoating are poorly infectious. The events that are occurring during uncoating are probably more complex than the simple "fate of capsid" assay might suggest. Suppressor mutations that restore normal infection to CA mutants do not always correct the capsid stability measured biochemically. While CA is clearly the key determinant of uncoating, it has been suggested that the viral virion infectivity factor (Vif) protein can affect the core stability.

Some host proteins have been implicated in either stabilizing or destabilizing the cores and so can impact the course of infection. One such protein is cyclophilin A, which is thought to stabilize the capsid and so promote infection especially in settings and genetic backgrounds where premature uncoating occurs. Another such protein is PDZD8, which binds to the HIV-1 capsid. Knockdown of PDZD8 destabilizes capsids and inhibits reverse transcription. Its mode of action likely involves the cytoskeleton. Other such host proteins include CPSF6 and TRIM5α.

Many CA mutants affect the efficiency of reverse transcription, suggesting that a proper core particle is needed for this process. These mutants support early observations of the so-called "endogenous reverse transcription reactions," carried out with purified virions, which demonstrate a requirement for intact particles for synthesis of complete viral DNA. Conversely, reverse transcription affects the course of uncoating. Studies of the RTC during the course of reverse transcription carried out in the endogenous RT reactions have shown changes in the particle morphology as DNA synthesis proceeds. These observations suggest that the complex events of reverse transcription may rearrange many of the proteins inside the core, and the DNA product itself may induce stresses on the core that promote its disassembly.

The course of uncoating as monitored by the state of the CA protein has been linked to the course of nuclear entry. While much of the CA protein is lost in the cytoplasm, these studies suggest that there may be more extensive removal of CA on nuclear entry. CA is likely involved in the entry process itself. Several CA mutants carry out normal reverse transcription but are defective in the later steps of nuclear entry and DNA integration. CA can interact directly with the NUP358 component of the nuclear pore, and structural studies raise the intriguing notion that its proline isomerase activity may induce conformational changes in CA that could be involved in nuclear import of the PIC.

The multimeric state of CA protein in the core is probably essential for its interaction with many host factors and small molecule inhibitors. Large surfaces of CA that span more than one monomer, as are present in the CA hexamer, provide the targets for the interactions with such proteins as Fv1, TRIM5α, CPSF6 and the nuclear pore protein NUP153, and the antiviral compounds PF74 and BI-2. The binding sites for the proteins and these drugs overlap, and CA mutants often coordinately affect binding and sensitivity to both.

Virus–receptor Interactions

The initial event in the life cycle of a virus is its contact with the target cell. This encounter is mediated by interactions between components on the viral surface (capsid proteins or viral membrane anchored proteins) and several elements on the cellular membrane that can get attached to the virus. Most viruses establish contacts with cellular receptors which could be sialic acid bound molecules or proteins on the cell surface. Frequently, the first virus–cell contact is mediated by adhesion receptors that allow reversible attachment of the virus to particular cells or organs, mainly by means of electrostatic interactions. This initial interaction could lack specificity and is employed to enhance the encounter between virus and entry receptors, which irreversibly bind the virus, allowing them access to the intracellular milieu either by direct endocytosis/pinocytosis or by inducing fusion or penetration of the viral genome into the cell.

Characterization of the virus–receptor interactions has been a focus of sustained interest and many outstanding contributions of high-resolution X-ray structures of viral proteins in complex with their specific receptors have been reported during the last few years. Haemagglutinin (HA)–receptor interaction was extensively studied in different influenza virus subtypes. Influenza HA is synthesized in the infected cell as a polypeptide precursor (HA0) of about 550 amino acids that is proteolytically cleaved to generate the HA1 (approximately the N-terminal two thirds) and HA2 (the C-terminal third) chains that remain covalently linked by a disulfide bond. At the newly created HA2 N-terminus there is a stretch of hydrophobic amino acids, called the fusion peptide, which is inserted into the membrane during fusion. The overall structure of the influenza HA resembles an elongated spike sticking out of the membrane. The distal head, formed exclusively by HA1 sequences, bears the receptor-binding site, formed by a shallow pocket exposed on an outward-forming surface. Different residues within this pocket mediate viral attachment to the α2,3- or α2,6-linked sialic acid moieties in avian or human cellular surfaces, respectively, as observed by X-ray crystallography in different complexes with receptor analogues. The sialic acid binding pocket of HA1 is formed by a loop–helix–loop structure composed of the 130-loop, 190-helix and 220-loop.

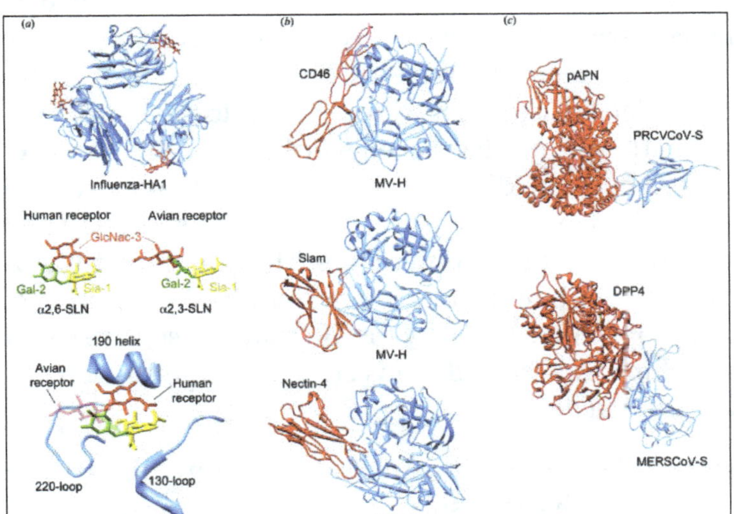

(a) The sialic acid binding pocket of influenza HA. A ribbon representation of an HA1 trimer is shown in the upper part. The structure of the bound human receptor analogue is shown as red sticks. The components of the human (left) and avian (right) receptor analogues are shown in the middle part of the figure and a close-up of the receptor-binding pocket with the superimposition of the two ligands is shown in the lower part. (b) The MV-H protein bound to the receptors: CD46, Slam and nectin-4. Ribbon drawings, showing the MV-H molecules in blue and the different receptor molecules depicted in red. (c) Structure of the PRCV CoV S-APN complex. (d) Structure of the complex MERS-CoV RBD-DPP4 complex.

A repertoire of crystal structures of HA–receptor complexes from different influenza outbreaks has recently been reported, providing a picture of the evolutionary path for the emergence of the H3N2 Influenza strain between 1968 and 2002. In particular, high-resolution X-ray structures have shown how the point mutation Asp225 to Asn in the 220-loop diminishes the specific HA–receptor interactions, justifying how this fixed mutation decreases the human receptor binding in H3N2 viruses circulating since the last decade, thereby reducing the impact of this influenza subtype. Moreover, the structures of different HA–receptor complex analogues have revealed the key determinants for animal-to-human transmissibility of the highly pathogenic avian H5N1: a single substitution (Gln to Leu) promotes binding to the human receptor to the detriment of the avian receptor. Interestingly, this mutation leading to increased hydrophobicity was responsible for the high transmissibility of human pandemic H2 and H3 influenza subtypes.

Recent crystallographic studies also revealed details of virus–host interactions for Measles virus (MV), which is a single-stranded, negative-sense, enveloped RNA virus. Three seminal works provide a landscape of atomic interactions between MV haemagglutinin (MV-H) and the protein receptors. The structure of MV-H bound to human CD46 fragment provided a rationale for the critical role of only two residues (Tyr481 and Gly546) in the alternative use of CD46 as a virus receptor, preserving the affinity for the primary SLAM receptor. The structure of the complex between MV-H and SLAM has also been determined. In addition, the structure of MV-H bound to a third receptor protein, nectin-4 which has been identified recently, has also been reported, showing that nectin-4 binds to MV-H by means of its N-terminal domain, establishing extensive hydrophobic interactions in the same concave lateral groove of MV-H as the other receptors but with minor differences. These high-resolution structures that illustrate the variability of interactions important for MV infection can also explain the extended virus tropism mediated by the alternative use of receptors or the receptor-specificity switching. Additionally, this structural information provides a framework for rational antiviral design.

Other crystallographic data have allowed characterization of the early steps of infection in important pathogens such as the single-stranded, positive-sense RNA Coronaviruses (CoV). The spike glycoprotein (S) of the CoV particle forms characteristic surface projections that are employed to interact with the target cell receptors. The determinants of CoV tropism are located in the distal globular domain of this viral membrane glycoprotein, which mediates attachment of the virus to the cell-surface molecules. In fact, the structure of the porcine respiratory coronavirus (PRCV) receptor-binding domain of the S protein in complex with its cell-receptor protein, the aminopeptidase N (pAPN), shows that the conformation of the receptor-binding edge in the envelope S protein contains the determinants of their receptor-binding specificity. This work also reports the interaction of a single glycan involved in protein interactions that must be conserved among a group of CoVs.

Very recently, Wang determined the structure of the receptor-binding domain (RBD) present in the S protein of the newly identified Middle East respiratory syndrome coronavirus (MERS-CoV) in complex with its cellular receptor, the dipeptidyl peptidase 4 (DPP4). Interestingly, this receptor does not resemble any other CoV receptor in either sequence or structure. On the other hand, the MERS-CoV RBD appears to be different from the SARS-CoV reported earlier, despite the relative similarity in the core subdomain, contributing to the variance in receptor specificities. The interphases of interactions consist of two regions distant from the DPP4 active site, where Li identified relevant residues that support the viral–receptor interaction that will potentially guide therapeutic strategies against this virus.

Viral and Eukaryotic Fusion Proteins

The viral fusion glycoproteins structurally characterized to date fall into three classes, although the membrane-fusion pathway seems to be very similar for all the enveloped viruses studied so far.

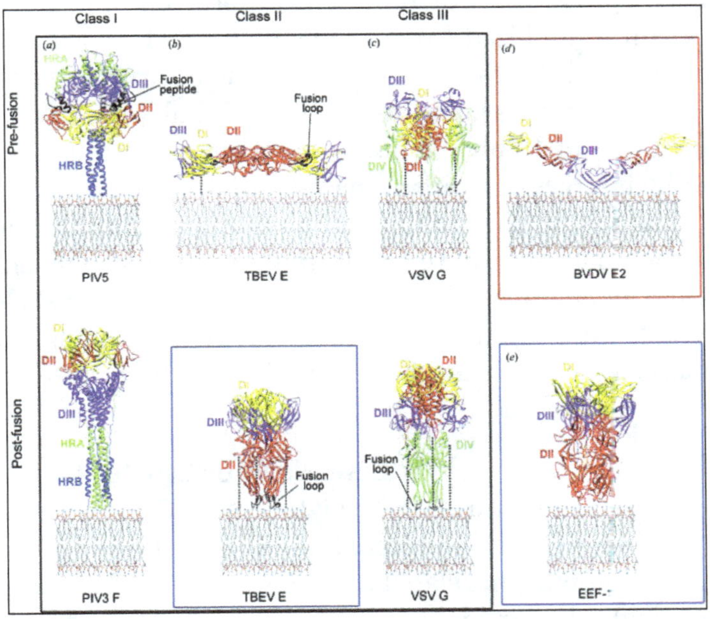

Class I fusogens are characterized by a seven-residue periodicity of nonpolar amino acids (called 'heptad repeats') that give rise to a central parallel trimeric α-helical coiled coil along the long axis of a rod-shaped molecule. The best characterized members of this class are the influenza virus HA and the fusion protein (F) of paramyxoviruses, but this class also includes fusion proteins from retroviruses, coronaviruses and filoviruses. Both the pre-fusion and post-fusion structures of class I fusion glycoproteins are trimeric. In most of the cases, the subunits constituting the trimer result from the proteolytic cleavage of a precursor into two fragments. The resulting C-terminal fragment, which is anchored in the viral membrane by a hydrophobic transmembrane (TM) domain, bears a hydrophobic fusion peptide at or near its amino-terminal end that is buried at a protein–protein interface in the pre-fusion state. This peptide gets inserted into the target membrane during fusion. Thus, in the post-fusion conformation, the shape of the trimeric molecule resembles an elongated hairpin-like structure with the fusion peptide and the TM domain located at the same end, as expected at the end of the fusion process.

The three classes of viral fusion glycoproteins. Ribbon diagrams of the pre-fusion (top) and post-fusion (bottom) conformations of the paramyxovirus (a); flavivirus (b); and rhabdovirus (c) proteins. For each class of fusogen, the equivalent protein domains are highlighted with identical colours and explicitly labelled (DI, yellow; DII, red; DIII, purple; DIV and heptad repeat region A, HRA, light green; HRB, blue; fusion peptide or loops, black). (d) The novel fold of the envelope protein E2 from BVDV. (e) The structure of the eukaryotic fusogen EEF1.

Until recently, class II proteins had only been found in flaviviruses (protein E) and alphaviruses (protein E1), which share many key characteristics. The first X-ray structure of a class II glycoprotein was that of the tick-borne encephalitis (TBE) flavivirus E protein ectodomain, solubilized from virions by limited trypsin digestion. Similar structures are now available for the ectodomain of denguevirus types 2 and 3. The polypeptide chain of the E protein is organized in three globular domains, essentially constituted by β-sheets. The hydrophobic fusion loop in the virion is located at the tip of domain II, buried by interactions with domain III of the adjacent monomer in the EE dimer. These proteins are attached to the viral membrane via a C-terminal TM anchor, which is linked by a flexible 'stem' region to the ectodomain. The envelope proteins from flaviviruses and alphaviruses assemble into icosahedral outer shells, but the mode of assembly differs in the two families, with alphaviruses forming canonical (T = 4) quasi-equivalent assemblies and flaviviruses forming unusual non-equivalent icosahedral assemblies.

Recent data show that the class II fold is more widely distributed than previously anticipated. Indeed, the crystal structures of the glycoprotein C (Gc) from Rift Valley fever virus (RVFV) reveal a class II fusion protein fold. RVFV belongs to the phlebovirus genus in the Bunyaviridae family, unrelated to flaviviruses or alphaviruses. The structure of RVFV Gc is strikingly similar to flavivirus E structures. In particular, both viruses share the same head-to-tail configuration of the protein dimers, with the fusion loop buried at the dimer interface. Also, the two fusion loops have the same tightly folded glycine-rich structure suggesting that phleboviruses may be evolutionarily related to alphaviruses and flaviviruses.

In another recent advance, the E1 protein of rubella virus, which belongs to the same Togaviridae family as alphaviruses, was found to have a class II fold, although with a more divergent structure that shows important differences in the fusion loops. In addition, rubella E1 does not form icosahedral assemblies.

Very recently, two independent groups have solved the structure of the larger envelope protein, E2, from the pestivirus bovine viral diarrhea virus (BVDV), another member of the Flaviviridae family. The structure unveiled that E2 has marked differences with the rest of the class II fusion proteins, defining a new structural class of fusogens.

Class III proteins are trimeric before and after fusion and share structural characteristics with both class I and class II fusion glycoproteins. A class III fold was identified in the structures of the ectodomains of the fusion glycoproteins G of vesicular stomatitis virus (VSV) and B (gB) of herpes simplex virus 1 (HSV-1). Their comparison revealed an unanticipated structural similarity between the two proteins, although no sequence similarity had previously been detected. The structures of the Epstein–Barr Virus (EBV) gB and baculovirus gp64 fusogens also belong to this class. VSV G is the only class III fusion protein for which the X-ray structures of both the pre-fusion and post-fusion states have been determined. The VSV G protein possesses both receptor-binding and fusion-promoter activities. As in the case of influenza virus, binding of rabdovirus G to a poorly characterized receptor at the cell surface induces endocytosis of the virus particle. Acidification of the endosome triggers G for membrane fusion. However, in contrast to all other fusion proteins, the low-pH inactivation of VSV G is reversible. The crystal structure of the G protein in the pre-fusion conformation revealed two fusion loops reminiscent of class II proteins that are oriented downwards towards the viral membrane. After low-pH exposure, the fusion domain moves upwards by flipping relative to the central core of the trimer to form an intermediate pre-hairpin structure. This is followed by the reversal of the molecule around the central rigid block formed by lengthening of the central helix in the grooves of the central core in an antiparallel manner. This six-helix bundle has an obvious resemblance to those of the class I proteins. It is likely that the transition of VSV G from the pre-fusion to the post-fusion conformation involves disassembly of the trimer into the monomers and reassembly into trimers upon interaction of the fusion loops with the target membrane.

While vesicle fusion is required for a number of essential biological processes such as exocytosis and synaptic transmission, cell–cell fusion processes are crucial in development. In all cases, membrane fusion follows the same steps already described for virus–host interaction. In analogy with virus–cell fusion, vesicle and cell–cell fusion requires the formation of highly stable protein assemblies that provide the energy necessary to overcome the repulsive forces of membranes in close proximity. Also vesicle and cell–cell fusion, as in viral fusion, require higher-order multimerization of the fusion proteins. The main difference between virus–cell fusion and vesicle or cell–cell fusion is that in the former process the protein fusogen is present only in the viral membrane. In contrast, the proteins involved in vesicle fusion and cell–cell fusion are initially inserted in the two membranes predestined to fuse. Studies of intracellular fusion events have revealed two families of fusion proteins, the SNARE [soluble N-ethylmaleimide-sensitive factor (NSF) attachment protein receptors] and the dynamin-like 'atlastin' GTPases. In both cases, membrane merger results from trans-oligomerization of molecules anchored in the opposed membranes, followed by a conformational change that pulls the two membranes towards each other. In contrast to vesicle fusion, cell–cell fusion entails the same set of fusion proteins in the two membranes. For instance, the exceptional process of hypodermal cell fusion in C. elegans to form a large multi-nucleated syncytium of all skin cells is driven by the epithelial fusion failure 1 protein, EFF-1. Unlike SNAREs and viral fusogens, EFF-1 has been shown to be required in both cell membranes for fusion. Recently, the 2.6 Å resolution X-ray structure of the EFF-1 protein has been determined by Felix Rey and colleagues. The EFF-1 trimer shows the same three-dimensional fold and quaternary conformation of post-fusion class II viral fusion proteins, although it lacks the nonpolar 'fusion loop', indicating that it does not insert into the target membrane and suggesting that membrane fusion driven by EFF-1 entails trans-trimerization. The blocking of EFF-1 trimerization interferes with the fusion reaction. The study also provides evidence that the monomeric form of EFF-1 is metastable and that trimerization is irreversible, matching the properties of the pre- and post-fusion forms, respectively, of the viral counterparts, demonstrating an evolutionary link with viral fusion proteins.

Genome Delivery in Bacteriophages

Bacteriophages show widely diverse structures and types of nucleic acid genomes; they have helical or icosahedral capsids and may not include a lipid envelope. Phages range in structural complexity from very small and simple non-enveloped icosahedral viruses (e.g. the ssDNA phage ϕX174) and long but simple helical viruses (filamentous phages), to large tailed viruses (e.g. the dsDNA phages T4 or ϕ29).

Bacteriophage tails are fascinating molecular machines specifically evolved to recognize host cells, penetrate the cell envelope barrier and deliver DNA into the cytoplasm. The Caudovirales tails display very different sizes and morphologies with lengths ranging from ~100 Å in some podophages to ~8000 Å in some siphoviruses. Long tails of Siphoviridae and Myoviridae phages consist of the tail tip complex, which is responsible for host recognition and initiation of the infection process, the tail tube, which makes a conduit for genomic DNA, and terminator proteins, which terminate the tail assembly and create the binding interface for head attachment. The tail of Myoviridae phages also contains a contractile sheath surrounding the tail tube. The tail tip complex has different size and morphology in different phages. Phages which use protein receptors for cell binding usually have conical tail tips e.g., SPP1 or λ, whereas phages using polysaccharide receptors usually have elaborate baseplates at the distal end of the tail e.g., T4 or the lactococcal phages TP901-1 and p2. Furthermore, phages usually have side-tail fibres or spikes, attached to the periphery of the tail tip complex, as well as a central tail spike.

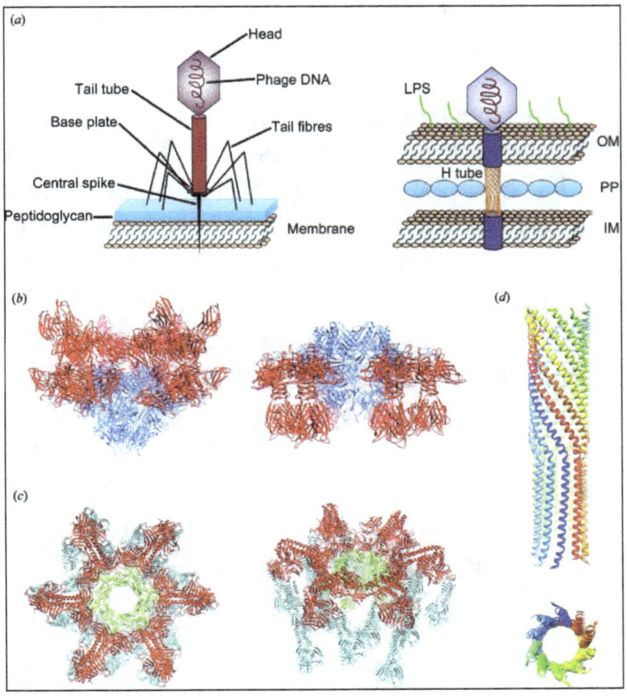

(a) Schematic representation of a tailed phage. LPS, lipopolysacharide; OM, outer membrane; PP, periplasmic space; IM, inner membrane. (b) The baseplate of phage P2 before activation, with the receptor-binding domains of the receptor-binding protein (red) pointing upwards, away from the bacterium. The movement of 200° by the receptor-binding protein that completely reverses the orientation. The right panel shows the baseplate of P2 after activation by calcium ions with the receptor-binding domains pointing downwards, towards the bacterium. (c) The baseplate of phage TP901-1, top view looking down the phage tail tube axis (left) and lateral view (right). (d) The structure of the H protein coiled-coil tube. Ribbon representation of the structure, with the monomers individually coloured. (bottom) Top view of the tube, with the N-terminus pointing to the viewer.

Bacteriophage baseplates vary in size and complexity. The most extensively studied Myoviridae baseplate is that of phage T4, which is composed of ~140 polypeptide chains of at least 16 different proteins. The structures of T4 baseplate before and after tail contraction have been determined using cryo-electron microscopy and the structures of nine baseplate proteins have been determined using X-ray crystallography. The T4 baseplate attaches six long tail fibres of ~1450 Å length, which reversibly bind to the Escherichia coli lipopolysaccharide (LPS) and/or OmpC molecules, serving for primary host recognition. It also has six short tail fibres attached to its periphery. Upon binding of the long tail fibres, a signal is transmitted to the baseplate causing the six short tail fibres to extend and bind irreversibly to the LPS, resulting in a conformational rearrangement where the baseplate switches from a dome-shaped form to a star-shaped conformation which, in turn, triggers the contraction of the tail sheath causing the specialized tip of the inner tail tube to puncture the outer membrane. The genome then passes through the tail tube into the cytoplasm.

Recent crystallographic and electron microscopy studies have revealed the structures of two Siphoviridae baseplates, those of phages p2 and TP901-1, infecting Gram-positive Lactococcus lactis. The baseplate of phage P2 is composed of three protein species. The central part of the baseplate is formed by a circular hexamer of ORF15 proteins with a central hole. A trimer of ORF16 is located at the bottom of the baseplate, forming a closed dome that does not allow DNA passage. Six ORF18 trimers are attached to the central ring, each trimer interacting with a carboxy-terminal extension of an ORF16 monomer. ORF18 is the receptor-binding protein (RBP) and electron microscopy reconstructions show that these proteins point upwards, towards the capsid in the free virion. In the presence of Ca^{2+}, a cation essential for infection, the RBP complex is rotated 200° to point downwards towards the host cell. This conformational change of the baseplate also leads to the separation of three ORF16 monomers, opening up a hole in the centre of the baseplate, presumably allowing passage of DNA into the host.

The baseplate of phage TP901-1 is composed of multiple copies of four different proteins. The centre is a hexameric circular core formed by the Dit protein. From the core, six arms emanate, each arm being composed of a trimer of the BppU protein. The arms are α-helical until the elbow. The rest of the arm points downwards and forms an adaptor domain for receptor-binding proteins. Three trimeric receptor-binding proteins bind to each adaptor leading to a total of 54 sites for binding the host cell envelope saccharides. The hole in the centre of the hexameric core is filled by a central tail fibre (Tal) as seen in the electron microscopy reconstruction. The receptor-binding proteins point downwards, i.e. towards the host bacterium, ready for adhesion. No Ca^{2+} ions are necessary for activation of TP901-1, suggesting that conformational changes are probably not needed for receptor binding. How receptor binding is related to DNA transfer in these phages is less clear; perhaps the strong binding with up to 54 receptor molecules pushes the central tail fibre against the cell wall and the force exerted by the cell wall against the tail is sensed by the other end of the tail fibre, which opens a hole at the end of the tail tube.

Very recently, Rossmann and colleagues have demonstrated that the small single-stranded (ss)DNA bacteriophage ϕX174 possesses a mechanism of DNA translocation similar to that described for the tailed phages. However, the ϕX174 'tail' protrudes from the virion only at the time of infection. ϕX174 is a small icosahedral microvirus with a circular ss(DNA). X-ray and electron microscopy studies showed that the virus capsid has spikes on all pentameric vertices. The 260 Å-diameter capsid is constructed from 60 copies of the F protein. In addition, each of the 12 spikes consists of five G proteins, protruding 32 Å above the F-protein shell. The capsid also contained 60 copies of the DNA-binding protein J and ten to 12 copies of the DNA pilot protein H. Until now, the structure and location of the H proteins remained unknown as all structure determinations of this phage made use of the icosahedral symmetry. Rossmann and colleagues have now solved the crystal structure of the central domain of the H protein at 2.4 Å resolution. It consists of a 170 Å-long α-helical tube built from ten α-helices. Each monomer is kinked at approximately the middle of the molecule, resulting in a tube with their amino termini arranged in a right-handed super-helical coiled coil and their carboxy termini arrayed in a left-handed super-helical coiled coil. The N-terminal domain, which is slightly conical, has a minimum internal diameter of approximately 22 Å, whereas the internal diameter of the C-terminal domain is about 24 Å. With a minimum internal diameter of 22 Å, the H tube can easily accommodate two unpaired ssDNA strands with intercalated bases in a similar way to that in which the circular ssDNA genomes of filamentous bacteriophages are packaged into cylindrical fivefold symmetric capsids with comparable inner dimensions. The H tube also appears long enough to span the periplasmic space, with both N- and C-terminal regions rich in Ala, Gly and Ser residues, which have high occurrence in transmembrane helices. Therefore, these regions would probably anchor the tube in the E. coli inner and outer membranes.

RNA-dependent RNA Polymerases

RNA-dependent RNA polymerases (RdRPs) are the catalytic components of the RNA replication and transcription machineries and the central players in the life cycle of RNA viruses. RdRPs belong to the superfamily of template-directed nucleic acid polymerases, including DNA-dependent DNA polymerases (DdDP), DNA-dependent RNA polymerases and reverse transcriptases (RT). All theses enzymes share a cupped right-hand structure, including fingers, palm and thumb domains, and catalyse phosphodiester bond formation through a conserved two-metal-ion mechanism. A structural feature unique to RdRPs is the 'closed-hand' conformation, as opposed to the 'open-hand' found in other polynucleotide polymerases. This closed-hand conformation is accomplished by interconnecting the fingers and thumb domains through the N-terminal portion of the protein and several loops protruding from fingers, named the fingertips, that completely encircle the active site of the enzyme. In the prototypic RdRPs, the closed 'right-hand' architecture encircles seven conserved structural motifs that are arranged in the order G, F, A, B, C, D and E containing a number of highly conserved amino acids. The only exception to this scheme is found in Birnavirus and Permutatetravirus polymerases, where motif C is encoded upstream of motif A. Each of the seven motifs in the RNA polymerase domain adopts a specific fold that extends beyond the regions of sequence similarity into the so-called homomorphs for most RNA virus RNA polymerases. Three well defined channels have been identified in the RdRP structures, serving as: the entry path for template (template channel) and for nucleoside triphosphates (NTP channel) and the exit path for the dsRNA product (central channel).

Overal structure of a viral RdRP. (a) Ribbon representation of a typical RdRP. The seven conserved motifs are indicated in different colours: motif A, red; motif B, green; motif C, yellow; motif D, purple; motif E, cyan; motif F, blue; motif G, pink. The side chains of the catalytic Asp residues in the active site are also shown as sticks. (b) Lateral view of a surface representation of the enzyme (grey) that has been cut to expose the three channels that are the entry and exit sites of the different substrates and reaction products. The structural elements that support motifs A, B and C are also shown as ribbons. (c) Sequential structures illustrating the movement of the different residues within the palm domain from a binary RdRP-RNA open complex (left) to an RdRP-RNA-rNTP open ternary complex (middle), and a closed ternary complex (right). Image based on different poliovirus elongation complexes. The different structures correspond to the 3D-RNA, 3D-RNA-CTP open complex and 3D-RNA-CTP closed complex structures of poliovirus elongation complexes, respectively.

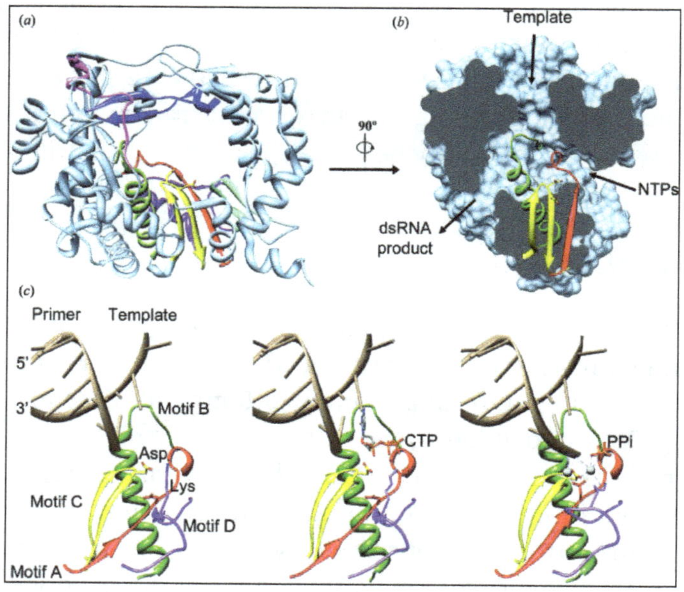

The recent explosion of structural and functional information on viral RdRPs has provided insights into both initiation of RNA synthesis and the elongation process. Correct initiation of RNA synthesis is essential for the integrity of the viral genome. There are two main mechanisms by which viral replication can be initiated: primer-independent or de novo, and primer-dependent initiation, reviewed in van Dijk. Briefly, in the de novo synthesis, one initiation nucleotide provides the 3′-hydroxyl for the addition of the next nucleotide whereas the primer-dependent initiation requires the use of either an oligonucleotide or a protein primer as provider of the hydroxyl nucleophile. The members of the Picornaviridae and Caliciviridae families use exclusively the protein-primed mechanism of initiation. In this process, a tyrosine residue provides the hydroxyl group for the formation of a phosphodiester bond with the first nucleotide. Enzymes using primer-dependent initiation display a more accessible active-site cavity, enabling them to accommodate the small VPg protein that acts as a primer protein in RNA synthesis. The foot-and-mouth disease virus (FMDV) VPg protein lines the RNA-binding cleft of the corresponding RdRP, positioning its Tyr3 OH group as a molecular mimic of the free 3′-OH group of a nucleic acid primer at the active site for nucleotidylylation, thereby initiating replication. In the presence of oligoadenylate and UTP, the product of the reaction, VPg–UMP, can be observed in the crystal structure, at a position remarkably similar to the position of the template–primer RNA duplex. After nucleotidylylation of VPg, some structural rearrangements of the RdRP follow, marking the transition from initiation to the elongation phase of RNA synthesis. By contrast, enzymes using the de novo initiation, Reovirus and Flavivirus RdRPs, contain additional structural elements that fill most of the active-site cavity, providing a support platform for the primer nucleotides, thereby enabling de novo initiation of RNA synthesis. Crucially, these protrusions are also able to undergo large conformational rearrangements to facilitate translocation of the RNA recently synthesized.

The replication elongation process can be roughly divided into three steps, including nucleotide selection, phosphodiester bond formation and translocation to the next nucleotide for the subsequent round of nucleotide addition. The structures of a large number of RdRP-RNA-rNTP replication–elongation complexes have been determined for different members of the Caliciviridae and Picornaviridae families. These studies have provided important insight into the structural alterations associated with each catalytic step. These structures indicated that RdRPs use subtle rearrangements within the palm domain to fully configure the active site for catalysis upon correct rNTP binding. Among these movements, a flexible loop at the N-terminus of motif B assists in the correct positioning of the template nucleotide in the active site, facilitating the binding of the incoming rNTP via base pairing to a fully prepositioned templating base and stacking with the primer. Binding of the correct nucleotide then induces the realignment of β-strands in the palm subdomain that includes structural motifs A and C, resulting in the repositioning of the catalytic aspartate in motif A to allow interactions with both the metal ions required for RdRP function. Recent data also suggest that a conserved lysine residue within motif D can coordinate the export of the inorganic phosphate (PPi) group from the active site, once catalysis has taken place, thereby triggering the end of the reaction cycle and allowing translocation of RNA. Finally, structural and functional data in enteroviruses indicate that steric clashes between the motif-B loop and the template RNA would also promote translocation.

Furthermore, very recently Ng and colleagues provided evidence for conformational changes occurring in the calicivirus RdRPs during catalysis. These structural comparisons revealed that, in addition to the active-site closure, the NV RdRP exhibits two other key changes: a rotation of the central helix in the thumb domain by 22°, resulting in the formation of a binding pocket for the primer RNA strand, and the displacement of the C-terminal tail region away from the central active-site groove, which also allows the rotation of the thumb helix.

Oligomerization of RdRPs has been reported for a number of Picornavirus, Flavivirus and Calicivirus enzymes. Oligomerization may lead to cooperative template binding, lattice formation, and a stimulation of viral RNA polymerase activity in vitro, which in Flavivirus appears to be specific for the initiation of RNA synthesis. Also inactive RdRPs can induce stimulation of activity and participate in array formation, which has led to the hypothesis that this mechanism has evolved to stabilize the enzyme during replication. The dimerization of the Hepatitis C virus (HCV) RdRp has been proposed to be mediated by the thumb domain, whereas in PV, residues in the N-terminal part of the polymerase domain and two interface I aspartates appear to be crucially involved.

RdRPs synthesize RNA using an RNA template. This biochemical activity, not present in mammalian cells, offers the opportunity to identify very selective inhibitors of this viral enzyme. Antiviral drugs targeting the RdRPs may either directly inhibit polymerase activity or essential interactions with the RNA template, or RdRP–RdRP contacts promoting oligomerization, or interactions with other regulatory proteins. The detailed structural and mechanistic understanding of the conformational changes that occur during catalysis is essential not only for understanding viral replication at the molecular level but also for the design of novel inhibitors capable of trapping the enzyme in specific conformational states. The Flaviviruses, Hepatitis C virus, Dengue virus and West Nile virus, as well as calicivirus NV are clear illustrations of the extent of efforts directed towards developing drugs that inhibit viral replication.

Viral Proteases

In most single-stranded RNA viruses, some double-stranded RNA viruses and retroviruses with polycistronic genomes, various functional protein domains are initially expressed as a single polyprotein. Cleavage of the polyprotein into individual functional units is essential for their survival. Hence, polyprotein processing is an integral step in the replication of these viruses. Extensive structural studies on proteases of animal viruses have been carried out as they are potential candidates for antiviral therapy. The PDB has 307 entries of viral proteases of which 276 are X-ray crystal structures. When filtered at 90% sequence uniqueness, the number of proteases is still as large as 93 of which 60 are structures with bound ligands. Viral proteases possess many unique features which distinguish them from cellular and bacterial proteases. These features include stringent specificity, regulation of activity, novel constellation of residues at the active site, amino-acid sequence unrelated to other proteases of the same class, and adaptability to multiple roles.

Viral proteases are very stringent in their substrate specificities, unlike cellular proteases whose specificities are normally dependent on the P1 residue only. In many viral proteases, the efficiency of cleavage at a particular site of the polyprotein is dependent on the sequence on either side of the cleavable bond and P1 to P4 and P1′, P2′, etc. are also important for specificity. The conformation and accessibility of the peptide bond also contribute to the susceptibility to cleavage. Each susceptible site in a polyprotein may have a different cleavage profile. This allows a way of controlling the cleavage events such that the proportions of individual domains are controlled and their activities regulated. The stringent substrate specificity of viral proteases has led to their application in molecular biology as a reagent for removal of affinity tags from fusion proteins. The most widely used viral protease for such applications is the tobacco etch virus NIa protease.

Many novel combinations of active-site residues have been identified among viral proteases. One such example is the serine-like cysteine protease family of viral proteases, which possesses a cysteine instead of a serine in the catalytic triad in a trypsin- or chymotrypsin-like fold. The crystal structure of the HRV-14 $3C^{pro}$ revealed the adjustments in the catalytic site required for accommodating a cysteine in the place of serine in the catalytic triad of serine-like cysteine proteases. The serine proteases of the cytomegalovirus family have a catalytic triad of histidine, histidine and serine. The members of the birnavirus family possess a protease with serine and lysine residues forming a catalytic dyad, similar to the Lon proteases in bacteria. The aspartic protease of HIV is also different compared to the cellular aspartic proteases in that it functions as an obligate dimer with the two catalytic aspartates contributed by two different subunits. A similar example of a composite active-site formation is the case of $NS2^{pro}$ of HCV, which has a catalytic triad of histidine, glutamate from one monomer and cysteine from another. The active-site geometry is similar to that in serine proteases. The existence of an altered catalytic triad might result in reduction of catalytic efficiency of viral proteases compared to cellular proteases. Such characteristics are suited to the viral life cycle since, instead of efficiency in proteolysis, viruses require greater stringency in specificity and regulation of function. In the cytomegalovirus protease with a novel catalytic triad, a slow proteolytic step may be optimal for viral assembly, as proteolysis should occur only after the capsid maturation, which may be a relatively slow process. Some of the viral proteases have a structural fold similar to serine proteases although there is no sequence similarity with cellular proteases. The Sesbania mosaic virus protease has a chymotrypsin-like fold and is shown to be similar to Bacillus intermedius glutamyl endopeptidase. Athough the catalytic triad can be superposed well with serine proteases, the protease by itself is inactive.

The catalytic function of viral proteases has been found to be regulated by other domains that may be attached covalently to the N- or C-terminus of the protease. These regulatory mechanisms also serve to control the time and place of proteolytic activity. It has been found that NS2A and NS4A act as cofactors for the regulation of activity of the NS3 proteases of dengue virus and hepatitis C virus, respectively. These cofactors also contain membrane-spanning domains, which might co-localize the proteases to the membrane before activation. The oligomerization status of the viral proteases also appears to be variable compared to their cellular counterparts. The aspartyl protease of HIV is an obligate dimer. It has been demonstrated that the interaction of Sesbania mosaic virus (SeMV) protease with VPg in cis-(Protease-VPg) leads to the activation of the protease. Similarly VPg covalently linked to the N-terminus of pepper vein banding virus (PVBV) protease NIaPro (VPg-NIaPro) results in the enhancement of activity. In both these cases, the interaction of the disordered VPg domain alters the structure and function of the interacting proteases. It was shown that the poliovirus 3CD precursor, which consists of the protease domain 3Cpro and the polymerase domain 3Dpol, also exhibits protease activity similar to 3Cpro, but the polymerase is active only as 3Dpol after the elimination of the 3Cpro domain. Dengue virus protease shows 3000- to 6000-fold activation in the presence of the NS2B cofactor.

Another feature of viral proteases is their extreme sequence diversity, not only amongst themselves but also with respect to cellular proteases. No significant sequence identity has been found between viral proteases and cellular proteases, though the structural fold might be maintained. Most of the viral protease X-ray crystal structures have therefore been determined by ab initio methods of phase determination using isomorphous replacement or anomalous dispersion. In most cases, only the catalytic residues are conserved. The Coronaviral proteases contain an additional domain consisting mainly of α-helices at the C-terminus which is essential for dimerization. The high diversity of viral proteases might be due to the accelerated replication of the viruses and absence of a proof-reading mechanism in the replication of viruses with an RNA genome.

Many viral proteases perform more than one role. Hence, there are certain features in these proteases that confer unique sequence and structural properties compared to normal cellular proteases. The proteases of the alphaviruses have a compact structure resulting from shortening of loops such that their assembly into capsids is facilitated. RNA-binding motifs incorporated into the canonical protease domains are observed with many viral proteases of RNA viruses. In addition to the main function of polyprotein processing, viral proteases also appear to be involved in shutting down of host-cell translation by acting upon integral components of the host translation machinery. For example, the viral protease 2Apro encoded by the HRV genome cleaves the eukaryotic initiation factor eIF-4G of the host, preventing host protein synthesis. Sindbis core protein is not only a protease undergoing autocleavage from the polyprotein followed by self-inactivation, it also assembles to form the viral capsid. Helper component protease (HCpro) has been shown to act as a suppressor of post-transcriptional gene silencing. However, the protease function may be independent of suppressor activity. Antiviral drug discovery with proteases of the animal viruses as targets will be benefited by consideration of the special features of these proteases.

Capsid

A capsid is the protein shell of a virus. It consists of several oligomeric structural subunits made of protein called protomers. The observable 3-dimensional morphological subunits, which may or may not correspond to individual proteins, are called capsomeres. The capsid encloses the genetic material of the virus.

Capsids are broadly classified according to their structure. The majority of viruses have capsids with either helical or icosahedral structure. Some viruses, such as bacteriophages, have developed more complicated structures due to constraints of elasticity and electrostatics. The icosahedral shape, which has 20 equilateral triangular faces, approximates a sphere, while the helical shape resembles the shape of a spring, taking the space of a cylinder but not being a cylinder itself. The capsid faces may consist of one or more proteins. For example, the foot-and-mouth disease virus capsid has faces consisting of three proteins named VP1–3.

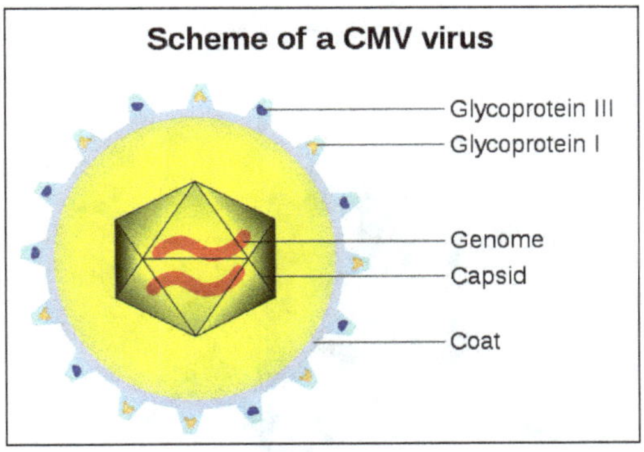

Schematic of a cytomegalovirus.

Some viruses are enveloped, meaning that the capsid is coated with a lipid membrane known as the viral envelope. The envelope is acquired by the capsid from an intracellular membrane in the virus' host; examples include the inner nuclear membrane, the golgi membrane, and the cell's outer membrane.

Once the virus has infected a cell and begins replicating itself, new capsid subunits are synthesized using the protein biosynthesis mechanism of the cell. In some viruses, including those with helical capsids and especially those with RNA genomes, the capsid proteins co-assemble with their genomes. In other viruses, especially more complex viruses with double-stranded DNA genomes, the capsid proteins assemble into empty precursor procapsids that includes a specialized portal structure at one vertex. Through this portal, viral DNA is translocated into the capsid.

Structural analyses of major capsid protein (MCP) architectures have been used to categorise viruses into lineages. For example, the bacteriophage PRD1, the algal virus Paramecium bursaria Chlorella virus (PBCV-1), mimivirus and the mammalian adenovirus have been placed in the same lineage, whereas tailed, double-stranded DNA bacteriophages (Caudovirales) and herpesvirus belong to a second lineage.

Specific Shapes

Icosahedral

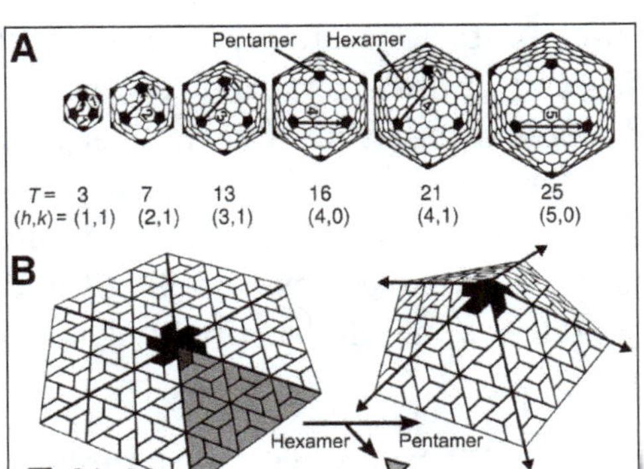

Virus capsid T-numbers.

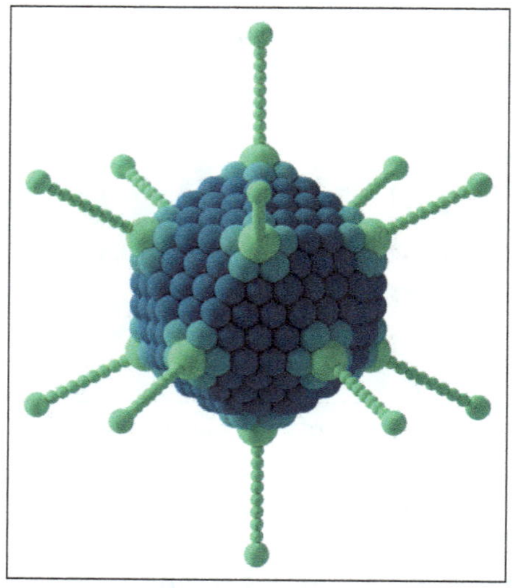

Icosahedral capsid of an adenovirus.

The icosahedral structure is extremely common among viruses. The icosahedron consists of 20 triangular faces delimited by 12 fivefold vertexes and consists of 60 asymmetric units. Thus, an icosahedral virus is made of 60N protein subunits. The number and arrangement of capsomeres in an icosahedral capsid can be classified using the "quasi-equivalence principle" proposed by Donald Caspar and Aaron Klug. Like the Goldberg polyhedra, an icosahedral structure can be regarded as being constructed from pentamers and hexamers. The structures can be indexed by two integers h and k, with $h \geq 1$ and $k \geq 0$; the structure can be thought of as taking h steps from the edge of a pentamer, turning 60 degrees counterclockwise, then taking k steps to get to the next pentamer. The triangulation number T for the capsid is defined as:

$$T = h^2 + h \cdot k + k^2$$

In this scheme, icosahedral capsids contain 12 pentamers plus 10(T – 1) hexamers. The T-number is representative of the size and complexity of the capsids. Geometric examples for many values of h, k, and T can be found at List of geodesic polyhedra and Goldberg polyhedra.

Many exceptions to this rule exist: For example, the polyomaviruses and papillomaviruses have pentamers instead of hexamers in hexavalent positions on a quasi-T=7 lattice. Members of the double-stranded RNA virus lineage, including reovirus, rotavirus and bacteriophage φ6 have capsids built of 120 copies of capsid protein, corresponding to a "T=2" capsid, or arguably a T=1 capsid with a dimer in the asymmetric unit. Similarly, many small viruses have a pseudo-T=3 (or P=3) capsid, which is organized according to a T=3 lattice, but with distinct polypeptides occupying the three quasi-equivalent positions.

T-numbers can be represented in different ways, for example T = 1 can only be represented as an icosahedron or a dodecahedron and, depending on the type of quasi-symmetry,

T = 3 can be presented as a truncated dodecahedron, an icosidodecahedron, or a truncated icosahedron and their respective duals a triakis icosahedron, a rhombic triacontahedron, or a pentakis dodecahedron.

Prolate

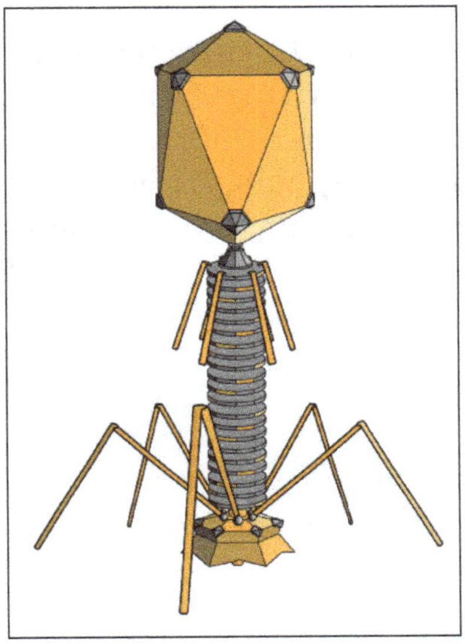

The prolate structure of a typical head on a bacteriophage.

An elongated icosahedron is a common shape for the heads of bacteriophages. Such a structure is composed of a cylinder with a cap at either end. The cylinder is composed of 10 elongated triangular faces. The Q number (or T_{mid}), which can be any positive integer, specifies the number of triangles, composed of asymmetric subunits, that make up the 10 triangles of the cylinder. The caps are classified by the T (or T_{end}) number.

Helical

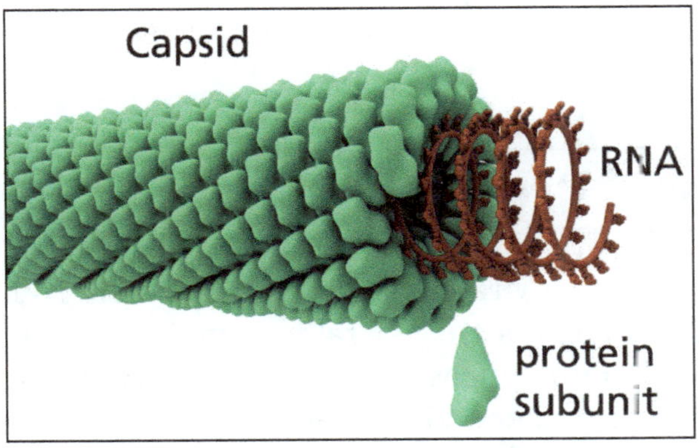

3D model of a helical capsid structure of a virus.

Many rod-shaped and filamentous plant viruses have capsids with helical symmetry. The helical structure can be described as a set of n 1-D molecular helices related by an n-fold axial symmetry. The helical transformation are classified into two categories: one-dimensional and two-dimensional helical systems. Creating an entire helical structure relies on a set of translational and rotational matrices which are coded in the protein data bank. Helical symmetry is given by the formula $P = \mu \times \rho$, where μ is the number of structural units per turn of the helix, ρ is the axial rise per unit and P is the pitch of the helix. The structure is said to be open due to the characteristic that any volume can be enclosed by varying the length of the helix. The most understood helical virus is the tobacco mosaic virus. The virus is a single molecule of (+) strand RNA. Each coat protein on the interior of the helix bind three nucleotides of the RNA genome. Influenza A viruses differ by comprising multiple ribonucleoproteins, the viral NP protein organizes the RNA into a helical structure. The size is also different the tobacco mosaic virus has a 16.33 protein subunits per helical turn, while the influenza A virus has a 28 amino acid tail loop.

Functions

The functions of the capsid are to:

- Protect the genome,
- Deliver the genome,
- Interact with the host.

The virus must assemble a stable, protective protein shell to protect the genome from lethal chemical and physical agents. These include forms of natural radiation, extremes of pH or temperature and proteolytic and nucleolytic enzymes. For non-enveloped viruses, the capsid itself may be involved in interaction with receptors on the host cell, leading to penetration of the host cell membrane and internalization of the capsid. Delivery of the genome occurs by subsequent uncoating or disassembly of the capsid and release of the genome into the cytoplasm, or by ejection of the genome through a specialized portal structure directly into the host cell nucleus.

Major Capsid Protein VP1

Major capsid protein VP1 is a viral protein that is the main component of the polyomavirus capsid. VP1 monomers are generally around 350 amino acids long and are capable of self-assembly into an icosahedral structure consisting of 360 VP1 molecules organized into 72 pentamers. VP1 molecules possess a surface binding site that interacts with sialic acids attached to glycans, including some gangliosides, on the surfaces of cells to initiate the process of viral infection. The VP1 protein, along with capsid components VP2 and VP3, is expressed from the "late region" of the circular viral genome.

Structure

VP1 is the major structural component of the polyomavirus icosahedral capsid, which has T=7 symmetry and a diameter of 40-45 nm. The capsid contains three proteins; VP1 is the primary component and forms a 360-unit outer capsid layer composed of 72 pentamers. The other two components, VP2 and VP3, have high sequence similarity to each other, with VP3 truncated at the N-terminus relative to VP2. VP2 and VP3 assemble inside the capsid in contact with VP1, with a stoichiometry of one VP2 or VP3 molecule to each pentamer. VP1 is capable of self-assembly into virus-like particles even in the absence of other viral components. This process requires bound calcium ions and the resulting particles are stabilized by, but do not require, inter-pentamer disulfide bonds.

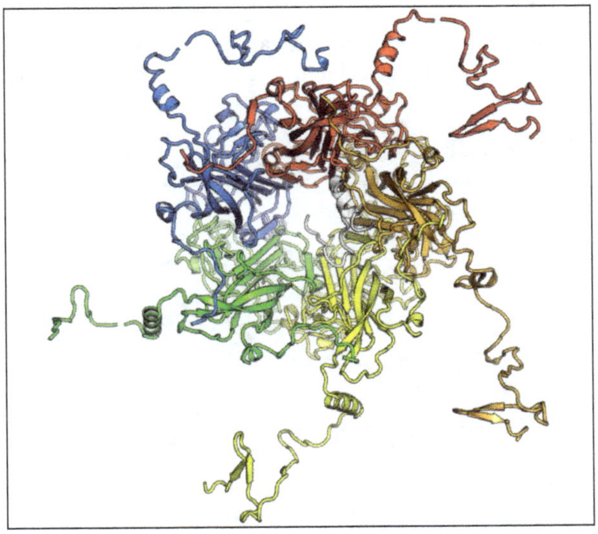

The structure of an individual pentamer of the murine polyomavirus VP1 protein. Each monomer is colored differently. The conformationally flexible C-terminal arms are shown here in conformations compatible with binding to neighboring molecules. Superposed is a fragment of the polyomavirus VP2 protein (white), which binds to a pentamer oriented toward the central cavity. VP1 is from PDB: 1SIE; VP2.

The VP1 protein monomer is primarily composed of beta sheets folded into a jelly roll fold. Interactions between VP1 molecules within a pentamer involve extensive binding surfaces, mediated in part by interactions between edge beta-strands. The VP1 C-terminus is disordered and forms interactions between neighboring pentamers in the assembled capsid. The flexibility of the C-terminal arm will enable it to adopt different conformations in the six distinct interaction environments imposed by the symmetry of the icosahedral assembly. The C-terminus also contains a basic nuclear localization sequence, while the N-terminus - which is oriented toward the center of the assembled capsid - contains basic residues that facilitate non-sequence-specific interactions with DNA.

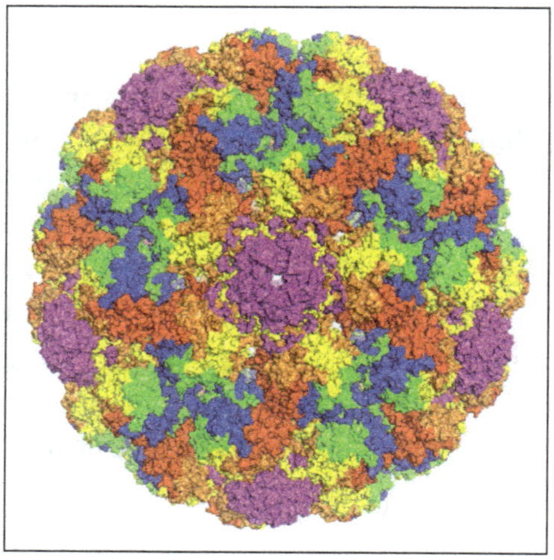

The same capsid structure as above, colored to illustrate the assembly of the icosahedral architecture from VP1 pentamers. Each symmetry-related VP1 monomer is shown in a different color.

Function and Trafficking

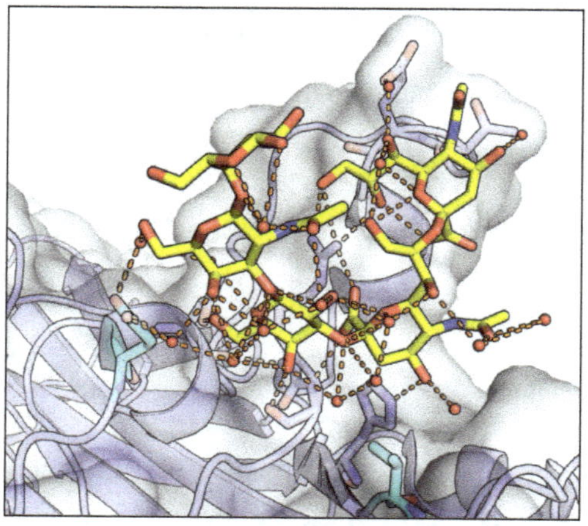

Murine polyomavirus VP1 in complex with the GT1a glycan. GT1a is shown in yellow and the VP1 monomer with a white surface and a blue protein backbone. A complex network of hydrogen bonds, many water-mediated, is shown at the binding surface by orange lines, with participating protein residues shown as sticks. Mutations of the two residues shown in cyan at the bottom of the figure can significantly affect pathogenicity.

The VP1 protein is responsible for initiating the process of infecting a cell by binding to sialic acids in glycans, including some gangliosides, on the cell surface. Canonically, VP1 interacts specifically with α(2,3)-linked and α(2,6)-linked sialic acids. In some cases additional factors are necessary conditions for viral entry; for example, JC virus requires the 5HT2A serotonin receptor for entry, although the specific mechanism of this requirement is unclear. Once attached to the cell surface, the virions enter the cell and are trafficked by a retrograde pathway to the endoplasmic reticulum. The exact mechanism of endocytosis varies depending on the virus, and some viruses use multiple mechanisms; caveolae-dependent mechanisms are common. The process by which polyomaviruses penetrate the membrane and exit the ER is not well understood, but conformational changes to VP1, possibly including reduction of its disulfide bonds, likely occur in the ER. For some polyomaviruses, VP1 has been detected reaching the nucleus along with the viral genome, though it is unclear how the genomic DNA disengages from VP1.

All of the capsid proteins are expressed from the late region of the viral genome, so named because expression occurs only late in the infection process. VP1 has a nuclear localization sequence that enables import from the cytoplasm where it is synthesized by the host translation machinery to the cell nucleus where new virions are assembled. This nuclear import process, mediated by karyopherins, acts on assembled VP1 pentamers in complex with VP2 or VP3; oligomerization to form capsids occurs in the nucleus.

Minor Capsid Proteins VP2 and VP3

Minor capsid protein VP2 and minor capsid protein VP3 are viral proteins that are components of the polyomavirus capsid. Polyomavirus capsids are composed of three proteins; the major component is major capsid protein VP1, which self-assembles into pentamers that in turn self-assemble into enclosed icosahedral structures. The minor components are VP2 and VP3, which bind in the interior of the capsid.

Gene Expression

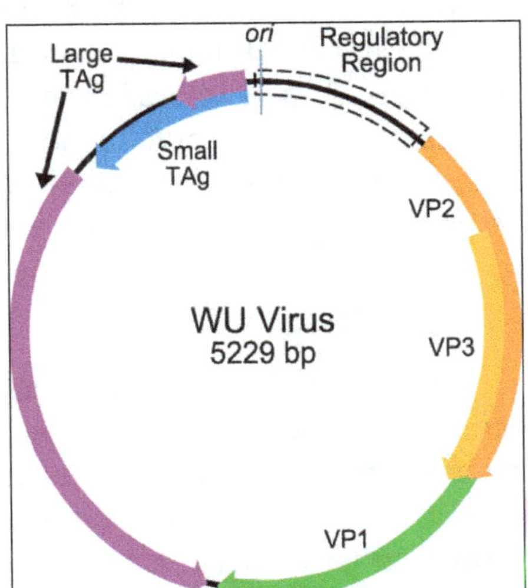

The circular genome of a representative polyomavirus, WU polyomavirus, with the late region at right indicating positions of the VP1, VP2, and VP3 genes.

All three capsid proteins are expressed from alternative start sites on a single transcript of the "late region" of the circular viral chromosome (so named because it is transcribed late in the process of viral infection). The VP3 start site is in frame downstream from that of VP2; in consequence VP3's sequence is identical to the C-terminal portion of VP2, which has an additional N-terminal extension. In at least some polyomaviruses, the VP2 N-terminus is myristoylated. Some members of the polyomavirus family, such as Merkel cell polyomavirus, do not appear to encode or express VP3, though VP2 is present.

Structure and Interactions

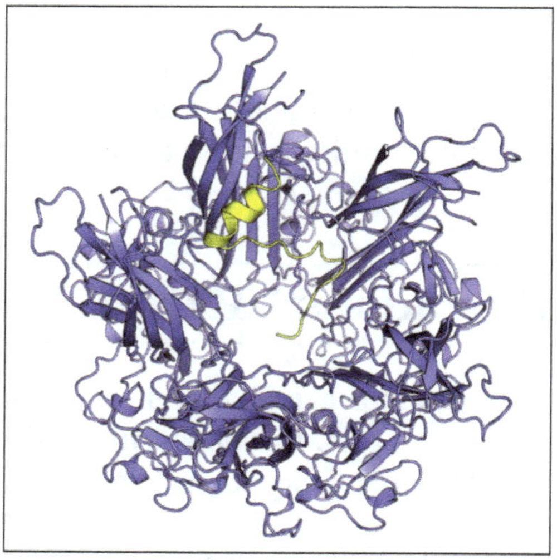

A fragment of the murine polyomavirus VP2 (yellow) in complex with the major capsid protein VP1 (blue).

Both VP2 and VP3 are primarily intrinsically unstructured proteins; they have DNA-binding domains and a nuclear localization signal at their C-terminal ends. Both VP2 and VP3 bind to the interior of VP1 pentamers in the assembled capsid. It is generally believed that the stoichiometry of this interaction is one molecule of VP2 or VP3 to each VP1 pentamer, though higher ratios have sometimes been reported, possibly indicating that pentamers can accommodate associations with two minor proteins.

Function

VP2 and VP3 are thought to be involved in facilitating viral entry into the host cell, either by mediating associations with and exit from the endoplasmic reticulum or by facilitating the entry of the viral genome into the cell nucleus. However, the precise mechanism of their involvement is unclear, and may vary among polyomaviruses. In most studies, viral propagation is either reduced or abrogated in the absence of one or both proteins, but the apparent mechanisms vary; for example, in JC virus both VP2 and VP3 seem to be essential for packaging the viral chromosome into the capsid, while absence of these proteins in SV40 prevents successful entry into new host cells, with variable effects on packaging reported. In Merkel cell polyomavirus, the effect of VP2 appears to vary depending on the cell type of the infected cell. In murine polyomavirus the minor proteins have been reported to induce apoptosis in the infected cell, and in SV40 they have been identified as viroporins.

Vpu Protein

Vpu is an accessory protein that in HIV is encoded by the vpu gene. Vpu stands for "Viral Protein U". The Vpu protein acts in the degradation of CD4 in the endoplasmic reticulum and in the enhancement of virion release from the plasma membrane of infected cells. Vpu induces the degradation of the CD4 viral receptor and therefore participates in the general downregulation of CD4 expression during the course of HIV infection. Vpu-mediated CD4 degradation is thought to prevent CD4-Env binding in the endoplasmic reticulum in order to facilitate proper Env assembly into virions. It is found in the membranes of infected cells, but not the virus particles themselves.

The Vpu gene is found exclusively in HIV-1 and some HIV-1-related simian immunodeficiency virus (SIV) isolates, such as SIV_{cpz}, SIV_{gsn}, and SIV_{mon}, but not in HIV-2 or the majority of SIV isolates. Structural similarities between Vpu and another small viral protein, M2, encoded by influenza A virus were first noted soon after the discovery of Vpu. Since then, Vpu has been shown to form cation-selective ion channels when expressed in Xenopus oocytes or mammalian cells and also when purified and reconstituted into planar lipid bilayers. Vpu also permeabilizes membranes of bacteria and mammalian cells to small molecules. Therefore, it is considered a member of the Viroporins family.

Expression

Vpu and Env are expressed from the same bicistronic mRNA in a Rev-dependent manner, presumably by leaky scanning of ribosomes through the vpu initiation codon. In fact Vpu gene overlaps at its 3′-end with the env gene. Several HIV-1 isolates were found to carry point mutations in the Vpu translation initiation codon but have otherwise intact vpu genes. Since removal of the Vpu initiation codon results in increased expression of the downstream env gene, it is possible that HIV-1 actually uses this mechanism as a molecular switch to regulate the relative expression of Vpu or Env in infected cells. The possible benefits of such a regulation are unclear.

Function

Two main functions have been assigned to the Vpu protein. The first function is known to induce degradation of the viral receptor molecule CD4, and the second function is to enhance the release of newly formed virions from the cell surface. Vpu accomplishes these two functions through two distinct mechanisms. In the case of CD4, Vpu acts as a molecular adaptor to connect CD4 to an E3 ubiquitin ligase complex resulting in CD4 degradation by cellular proteasomes. This requires signals located in Vpu's cytoplasmic domain. Enhancement of virus release on the other hand involves the neutralization of a cellular host factor, BST-2 (also known as CD317, HM1.24, or tetherin) and requires Vpu's TM domain. However, the exact mechanism of how Vpu counteracts BST-2 is still unclear. In the absence of Vpu, tetherin binds to the viral envelope and ties it to the cell membrane and other viral particles, impeding release of the viral particles. Recent data suggest that the BST-2 transmembrane domain is crucial for interference by Vpu. The interaction of Vpu and BST-2 results in the downregulation of BST-2 from the cell surface.

BST-2, which is an interferon (IFN)-inducible cell surface protein, appears to "tether" HIV to the cell in the absence of Vpu. BST-2 is a heavily glycosylated 29- to 33-kDa integral membrane protein with both a transmembrane domain and a putative glycosylphosphatidylinositol anchor (GPI). At the cell surface, BST2 resides in lipid rafts through the GPI anchor, whereas its TM domain lies outside them, indirectly interacting with the actin cytoskeleton. Vpu primary site of action is the plasma membrane, where this protein targets cell-surface BST-2 through their mutual TM-to-TM binding, leading to lysosomes, partially dependent on βTrCP.

Structure

Viral protein "u" (Vpu) is an oligomeric, 81-amino acid type I membrane protein (16 kDa) that is translated from vpu-env bicistronic mRNA. The N-terminus of Vpu encoding the transmembrane (TM) anchor represents an active domain important for the regulation of virus release but not CD4 degradation. The C-terminal cytoplasmic domain (54 residues) that contains a pair of serine residues (at positions 52 and 56) constitutively phosphorylated by casein kinase II. The phosphorylation of two serine residues in the cytoplasmic domain is critical for CD4 degradation in the ER. Based on 2D 1H NMR spectroscopy of a peptide corresponding to the cytoplasmic domain of Vpu, it was proposed that the cytoplasmic domain of Vpu contains two α-helical domains, helix-1 and helix-2, which are connected by an unstructured region containing the two conserved phosphoseryl residues. In addition, computer models predict a third α-helical domain in the transmembrane domain of Vpu, which could play an important role in the formation of ion channels.

Structural Viral Proteins

structural viral proteins make up the structure of a virus. Viruses contain a nucleic acid core, enclosed with a protein capsid. They may also have a lipid bilayer envelope; influenza A virus contains this envelope, while rhinovirus does not. Sometimes the envelope will also include glycoprotein spikes to help the virus attach to the host cell.

The protein capsid is made up of a strand of amino acids, forming a protein. This protein capsid determines the shape of the virus. The arrangement of individual capsomeres in the capsid is unique to each virus, and is used to identify the virus.

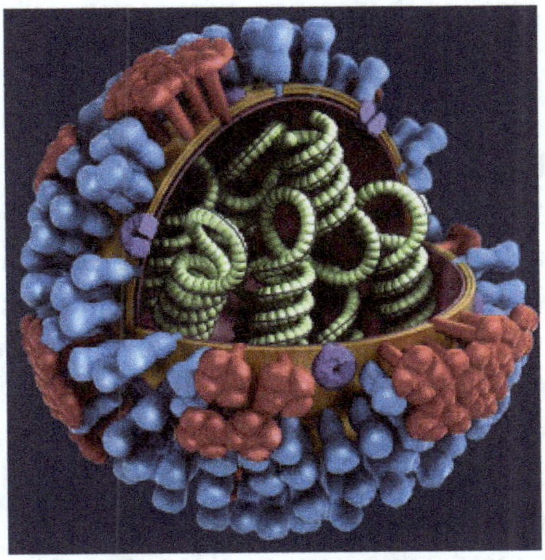

Influenza A has an envelope with spikes surrounding it to help it attach to the host cell.

Influenza A can be identified using the glycoprotein spikes on the virus. There are two kinds of spikes: hemagglutinin and neuraminidase. These spikes consist of a protein attached to a carbohydrate. They form a long stalk with a bulging head. These spikes can attach to the host cell. They're also what the immune system uses to identify the virus. It attacks these spikes, preventing the virus from attaching to the cell.

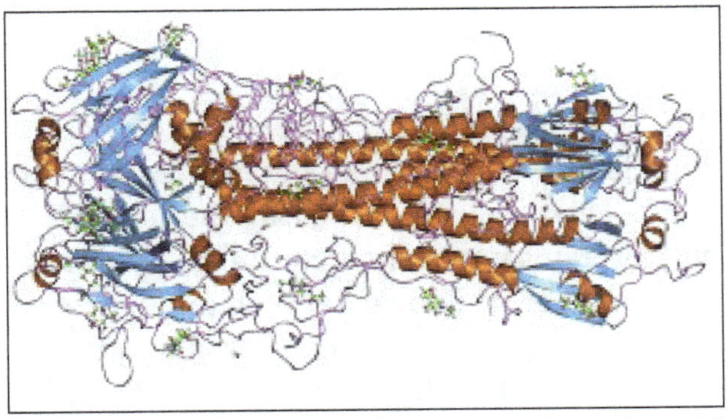

The glycoprotein hemagglutinin contains a long strand of proteins, with a large head.

Rhinovirus does not contain the envelope surrounding the protein capsid. There are around 100 individual rhinoviruses, and each is unique based on the individual proteins on the protein capsid, but they all form an icosahedral structure.

Ebola Viral Protein 24

Ebola viral protein 24 (eVP24) is considered a multifunctional secondary matrix protein present in viral particles. The broad roles eVP24 performs involve the formation of fully functional and infectious viral particles, promotion of filamentous nucleocapsid formation, mediation of host responses to infection, and suppression of the host innate immune system. It has been noted that eVP24 function can overlap with that of two other viral proteins; eVP40 matrix protein which functions in virus budding, and eVP35 which is also associated with immune suppression.

Function

eVP24 disrupts the signaling pathway of STAT1. The STAT1 protein is phosphorylated by interferons in response to viral infection causing it to express a non-classical nuclear localization signal and bind to the importin protein karyopherin-α (KPNA). Once bound to KPNA, STAT1 is transported to the nucleus where it stimulates gene transcription in response to viral infection. Classical nuclear localization signals are bound by arms 2-4 or by 6-8 of KPNA while non-classical nuclear localization signals are bound by KPNA 1, 5 and 6 in arms 8-10 allowing non classical signals to be transported at the same time as classical signals, providing faster signaling of certain signals. The eVP24 protein operates by binding to KPNA preventing the binding of STAT1. As a result, STAT1 is not able to elicit an immune response, however nuclear import is able to proceed as normal which may be important for viral replication. This means that eVP24 prevents the activation of an immune response against the Ebola virus without sacrificing its ability to have viral components transported to the nucleus or the target cell. eVP24 provides Ebola with an advantages over other viruses which disrupt STAT1 because unlike most other viruses, eVP24 uses mimicry of the STAT1 protein. This makes it very unlikely for the host to develop an adaptation as mutations in KPNA which prevent eVP24 binding are also likely to prevent STAT1 signalling.

Mechanism

eVP24 prevents the function of KPNA by binding in a region which overlaps with the binding region of STAT1. This is accomplished by the high binding affinity between KNPA and eVP24. These proteins have very high complementarity in the binding interface, similar to the complementarity shown between antibodies and antigens. In addition, the binding interface is large; over 2000 angstroms squared of solvent accessible surface is buried by the binding. Overall binding takes place with very little conformational change in either protein. There are three clusters of residues on eVP24 which form contacts to KPNA. These are located at residue positions 115 to 140, 184 to 186, and 201 to 207. Mutation of any single residue does not significantly reduce the binding of eVP24 to KPNA which demonstrates the robust mechanism of the viral protein. KPNA proteins have 10 armadillo repeats each consisting of three alpha helices which determine their binding specificity, the second helix from ARM 9 and 10 form a hydrophobic core with helix 6 of eVP24 adding to the stability of the complex. The binding sites for eVP24 and STAT1 have been shown to overlap. Mutation in any one of four residues in KPNA at positions 434, 474, 477 or 484 prevent binding to STAT1. Similarly, the mutations in residues 474, 477 or 484 of KPNA reduce the binding of eVP24. Additionally and critically, the binding of eVP24 does not prevent the binding of cargo proteins with classical nuclear localization signals as, like STAT1, eVP24 causes very little conformational changes in KPNA.

Effects on Symptom Progression

eVP24 acts as an antagonist to PY-STAT1 on KPNA. With eVP24 being transported to the nucleus instead of STAT1, the interferon-stimulated genes IFN-α/IFN-β and IFN-γ are disrupted and the cell does not enter an antiviral state. STAT1 has been shown to regulate the expression of certain immunoglobulins. More specifically, class switching from the predominant IgM to IgG2a was not present in STAT1-deficient cells. IgG2a plays a critical role in protection against pathogens and therefore without it, the cell is more susceptible to said pathogen. STAT1 has also been shown to regulate cell death by the inhibition of anti-apoptotic proteins Bcl-2 and Bcl-xL. STAT1 also induces the expression of procaspases, which are important factors in apoptosis signaling. When nuclear transport of STAT1 is inhibited pro-apoptotic signalling is disrupted, leading to decreased cell death.

Viral Matrix Protein

Viral matrix proteins are structural proteins linking the viral envelope with the virus core. They play a crucial role in virus assembly, and interact with the RNP complex as well as with the viral membrane. They are found in many enveloped viruses including paramyxoviruses, orthomyxoviruses, herpesviruses, retroviruses, filoviruses and other groups.

An example is the M1 protein of the influenza virus, showing affinity to the glycoproteins inserted in the host cell membrane on one side and affinity for the RNP complex molecules on the other side, which allows formation at the membrane of a complex made of the viral ribonucleoprotein at the inner side indirectly connected to the viral glycoproteins protruding from the membrane. This assembly complex will now bud out of the cell as new mature viruses.

Viral matrix proteins, like many other viral proteins, can exert different functions during the course of the infection. For example, in rhabdoviruses, binding of M proteins to nucleocapsids is accountable for the formation of its "bullet" shaped virions.

In herpesviruses, the viral matrix is usually called viral tegument and contains many proteins involved in viral entry, early gene expression and immune evasion.

M1 Protein

The M1 protein is a matrix protein of the influenza virus. It forms a coat inside the viral envelope. This is a bifunctional membrane/RNA-binding protein that mediates the encapsidation of RNA-nucleoprotein cores into the membrane envelope. It is therefore required that M1 binds both membrane and RNA simultaneously.

The M1 protein binds to the viral RNA. The binding is not specific to any RNA sequence, and is performed via a peptide sequence rich in basic amino acids.

It also has multiple regulatory functions, performed by interaction with the components of the host cell. The mechanisms regulated include a role in the export of the viral ribonucleoproteins from the host cell nucleus, inhibition of viral transcription, and a role in the virus assembly and budding. The protein was found to undergo phosphorylation in the host cell.

The M1 protein forms a layer under the patches of host cell membrane that are rich with the viral hemagglutinin, neuraminidase and M2 transmembrane proteins, and facilitates budding of the mature viruses.

M1 consists of two domains connected by a linker sequence. The N-terminal domain has a multi-helical structure that can be divided into two subdomains. The C-terminal domain also contains alpha-helical structure.

Virus-like Particle

Virus-like particles (VLPs) are molecules that closely resemble viruses, but are non-infectious because they contain no viral genetic material. They can be naturally occurring or synthesized through the individual expression of viral structural proteins, which can then self assemble into the virus-like structure. Combinations of structural capsid proteins from different viruses can be used to create recombinant VLPs. VLPs derived from the Hepatitis B virus (HBV) and composed of the small HBV derived surface antigen (HBsAg) were described in 1968 from patient sera. VLPs have been produced from components of a wide variety of virus families including Parvoviridae (e.g. adeno-associated virus), Retroviridae (e.g. HIV), Flaviviridae (e.g. Hepatitis C virus), Paramyxoviridae (e.g. Nipah) and bacteriophages (e.g. Qβ, AP205). VLPs can be produced in multiple cell culture systems including bacteria, mammalian cell lines, insect cell lines, yeast and plant cells.

VLPs can also refer to structures produced by some LTR retrotransposons (under Ortervirales) in nature. These are defective, immature virions, sometimes containing genetic material, that are generally non-infective due to the lack of a functional viral envelope. In addition, wasps produce polydnavirus vectors with pathogenic genes (but not core viral genes) or gene-less VLPs to help control their host.

Applications

Therapeutic and Imaging Agents

VLPs are a candidate delivery system for genes or other therapeutics. These drug delivery agents have been shown to effectively target cancer cells in vitro. It is hypothesized that VLPs may accumulate in tumor sites due to the enhanced permeability and retention effect, which could be useful for drug delivery or tumor imaging.

Vaccines

VLPs are useful as vaccines. VLPs contain repetitive, high density displays of viral surface proteins that present conformational viral epitopes that can elicit strong T cell and B cell immune responses. Since VLPs cannot replicate, they provide a safer alternative to attenuated viruses. VLPs were used to develop FDA-approved vaccines for Hepatitis B and human papillomavirus. More recently, VLPs were used to develop a pre-clinical vaccine against chikungunya virus.

Research suggests that VLP vaccines against influenza virus could provide stronger and longer-lasting protection against flu viruses than conventional vaccines. Production can begin as soon as the virus strain is sequenced and can take as little as 12 weeks, compared to 9 months for traditional vaccines. In early clinical trials, VLP vaccines for influenza appeared to provide complete protection against both the Influenza A virus subtype H5N1 and the 1918 flu pandemic. Novavax and Medicago Inc. have run clinical trials of their VLP flu vaccines.

Mycoviruses

Some fungi contain mycoviruses that lack the ability to be transmitted in cell free preparations and may be classified as VLPs. These are important in phytopathology, as they can cause hypovirulence in some species of phytopathogenic fungi.

Lipoparticle Technology

The VLP Lipoparticle was developed to aid the study of integral membrane proteins. Lipoparticles are stable, highly purified, homogeneous VLPs that are engineered to contain high concentrations of a conformationally intact membrane protein of interest. Integral Membrane proteins are involved in diverse biological functions and are targeted by nearly 50% of existing therapeutic drugs. However, because of their hydrophobic domains, membrane proteins are difficult to manipulate outside of living cells. Lipoparticles can incorporate a wide variety of structurally intact membrane proteins, including G protein-coupled receptors (GPCR)s, ion channels and viral Envelopes. Lipoparticles provide a platform for numerous applications including antibody screening, production of immunogens and ligand binding assays.

Expression Host Systems

The first step to creating a VLP is cloning and expressing the structural genes of interest. There are many systems to choose for expression. The chosen expression system can determine the limitations and effectiveness of the resulting VLP. The most well established expression systems being used today are as follows:

Bacterial systems are one of the most widely used, and are often based on the very well studied bacteria, Escherichia coli. This is a preferred method for production of recombinant proteins on a global scale due to the low cost and rapid nature of production, ease of scaling up, and high levels of expression. It is also possible to construct a VLP with multiple types of structural proteins. However, there are several disadvantages that come with use of this system:

- Inability to produce post-translational modifications.
- Inability to generate proper disulfide bonds within proteins.
- Other recombinant proteins of interest, particularly from eukaryotic cells, may be insoluble in an E. colisystem.
- Presence of endotoxins in generated proteins.

Research has suggested that culturing the cells at a low temperature or use of a fusion protein system can increase solubility for other proteins.

Yeast systems have been used to express structural genes of bacterial, yeast, plant and mammalian origin. Unlike bacterial systems, it is possible to introduce post-translational modifications and there is no endotoxin presence. Another disadvantage is that this system only allows for the creation of non-enveloped viruses. Yeast expression is unique in two ways: 1) to successfully produce a VLP using a yeast expression system, the bacteria must be propagated in bacteria before being introduced to the cell to create a stable transgene product and 2) Research has suggested that VLP assembly may occur more efficiently during the purification stage, instead of the cultivation stage. Pichia and Hansenulaare the most commonly used yeast strains.

Insect cell systems have fast growth rates in media without animal products, capacity for large scale cultivation, and the possibility of introducing post translational modifications. A baculovirus vectors always needs to successfully create the VLP. If more than one protein is required, the cell can be coinfected with a polycistronic vector, or infected with multiple monocistronic vectors. The latter method is preferred, because it allows for manipulation of individual protein levels, and identification of which ones are necessary. Although glycosylationis present, the patterns differ from that of mammalian cells, leading to a slightly different product.

Plant systems are less popular, but are good for the creation of VLPs with specific characteristics. Initially, plant-based expression systems gained popularity because they were attached to the idea of edible vaccines. It was thought that if an antigen was recombinantly expressed in a plant, ingestion of it would cause an immune response and effectively vaccinate the patient. Research has since moved away from edible vaccines for several reasons: administration of the vaccine by a medical professional is more likely to yield reproducible results, oral delivery was found to provide some protection against enteric pathogens, but not with any other body system, lack of antigen accumulation in the plant, and the avoidance of digestive acid and degrading enzymes.

The gene(s) for the protein(s) of interest are most commonly introduced using Agrobacterium. Once introduced, the gene can incorporate in either the nuclear or chloroplast genome. Although chloroplast transformation leads to very high copy numbers, it is a prokaryotic genome, so no glycosylation is observed. Genetic material can be introduced into the capsid during or after its assembly.

Mammalian systems are one of the most popular choices for researchers, making more than half of the recombinant proteins used in the pharmaceutical industry. While the complexity of construction and applications can often be a problem, it also leads to expression of highly efficient, high quality, complex VLPs that also have the correct glycosylation pattern. This system is useful for using a single polycistronic vector, as described above for the insect expression system. The recombinant VLPs are usually achieved using one of two methods:

- Adhesion Culture: Cells are seeded onto a surface and given proper nutrients.
- Suspension Culture: Cells are grown suspended in some type of culture media.

The latter method is more widely used when using mammalian cells to create VLPs.

Cell-Free Protein Systems (CFPS) are sometimes used to create VLPs. The following CFPS are commercially available: E. coli, wheat germs, insect cells, and rabbit reticulocytes.

Assembly

The understanding of self-assembly of VLPs was once based on viral assembly. This is rational as long as the VLP assembly takes place inside the host cell (in vivo), though the self-assembly event was found in vitro from the very beginning of the study about viral assembly. Study also reveals that in vitro assembly of VLPs competes with aggregation and certain mechanisms exist inside the cell to prevent the formation of aggregates while assembly is ongoing.

Linking Targeting Groups to VLP Surfaces

Attaching proteins, nucleic acids, or small molecules to the VLP surface, such as for targeting a specific cell type or for raising an immune response is useful. In some cases a protein of interest can be genetically fused to the viral coat protein. However, this approach sometimes leads to impaired VLP assembly and has limited utility if the targeting agent is not protein-based. An alternative is to assemble the VLP and then use chemical crosslinkers, reactive unnatural amino acids or SpyTag/SpyCatcher reaction in order to covalently attach the molecule of interest. This method has shown to be very effective at directing the immune response against the attached molecule, thereby inducing high levels of neutralizing antibody titres and breaking immune self-tolerance.

Purification of Non-enveloped VLPs

After the proteins of interest have been cloned and expressed in one of the above mentioned systems, they must be purified to get the final VLP product. Purification of non-enveloped VLPs generally involves four basic steps:

- Cell Lysis - cells are broken to release VLPs into solution.
- Cell Clarification - cellular debris is removed, leaving behind VLPs.
- Cell Concentration - The cell lysate (in this case, VLPs) are brought up to higher concentration in solution.
- Cell Polishing - removal of residual impurities.

These steps can be repeated multiple times in cycles depending on which protocol is used.

VP40

In molecular biology, VP40 is the name of a viral matrix protein. Most commonly it is found in the Ebola virus (EBOV), a type of non-segmented, negative-strand RNA virus. Ebola virus causes a severe and often fatal haemorrhagic fever in humans, known as Ebola virus disease. The virus matrix protein VP40 is a major structural protein that plays a central role in virus assembly and budding at the plasma membrane of infected cells. VP40 proteins work by associating with cellular membranes, interacting with the cytoplasmic tails of glycoproteins and binding to the ribonucleoprotein complex.

Structure

The VP40 monomer consists of two protein domains, the N-terminal oligomerization domain and the C-terminal membrane-binding domain, connected by a flexible linker. Both the N- and C-terminal domains fold into beta sandwich structures of similar topology. Within the N-terminal domain are two overlapping L-domains with the sequences PTAP and PPEY at residues 7 to 13, which are required for efficient budding. L-domains are thought to mediate their function in budding through their interaction with specific host cellular proteins, such as tsg101 and vps-4.

Function

VP40 coordinates numerous functions in the viral life cycle of the Ebola virus. These include: regulation of viral transcription, morphogenesis, packaging and budding of mature virions.

VP40 goes through intermediate states of assembly (e.g. octamers). It has been noted that proteins encoded by EBOV (VP30, VP35, and VP40) act independently as suppressors of RNA silencing, indicating that the virus actively resists cellular RNAi during replication.

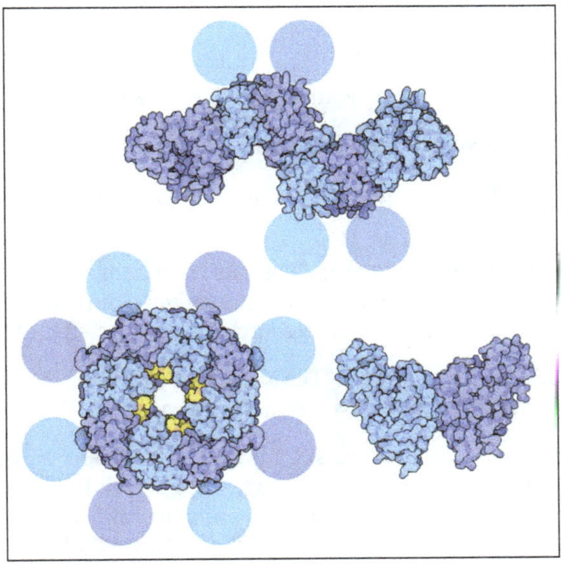

Different structures VP40 assumes in its functional roles.

Significance of VP40

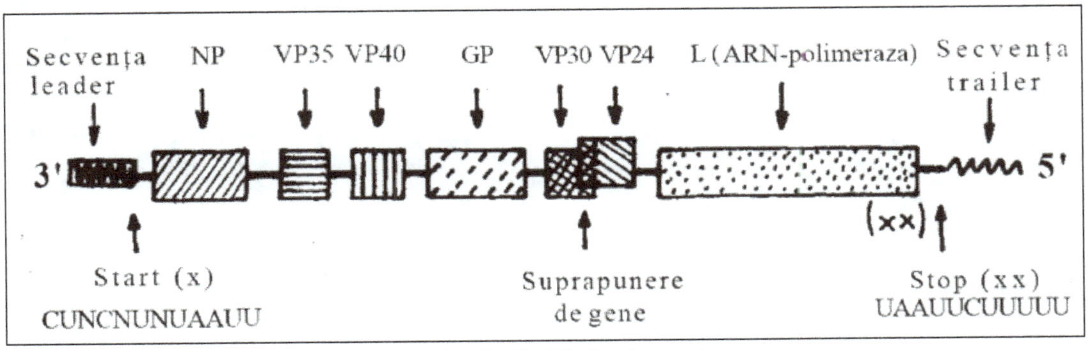

VP40 position in the Ebola virus genome.

Study of the matrix protein VP40 is important due to the high mortality rate of the Ebola virus, which is listed as a WHO Risk Group 4 Pathogen, an HHS Select Agent, an NIH/NIAID Category A Priority Pathogen, a CDC Category A Bioterrorism Agent, and a Biological Agent for Export Control by the Australia Group.

Expression of the matrix protein VP40 is sufficient to generate virus-like particles (no viral genetic material) in a mammalian host that are remarkably indistinguishable from live virus, from a morphological standpoint.

West African Ebola Virus Outbreak

During and after the 2014-2016 West African Ebola virus epidemic treatment options were sought; among them were the VP40 matrix protein as a target for possible research that may (or may not) lead to a therapeutic option.

References

- Viral-protein, topics-medicine-and-dentistry: sciencedirect.com, Retrieved 18 August, 2020
- Viral-protein-definition-components-functions: study.com, Retrieved 19 July, 2020
- Viral-nonstructural-proteins, immunology-and-microbiology: sciencedirect.com, Retrieved 10 April, 2020
- Ramqvist T, Dalianis T (February 2010). "Lessons from immune responses and vaccines against murine polyomavirus infection and polyomavirus-induced tumours potentially useful for studies on human polyomaviruses". Anticancer Research. 30 (2): 279–84. PMID 20332429
- Schowalter RM, Buck CB (2013-08-22). "The Merkel cell polyomavirus minor capsid protein". PLoS Pathogens. 9 (8): e1003558. doi:10.1371/journal.ppat.1003558. PMC 3749969. PMID 23990782
- Estrabaud E, Le Rouzic E, Lopez-Vergès S, Morel M, Belaïdouni N, Benarous R, Transy C, Berlioz-Torrent C, Margottin-Goguet F (July 2007). "Regulated degradation of the HIV-1 Vpu protein through a betaTrCP-independent pathway limits the release of viral particles". PLoS Pathog. 3 (7): e104. doi:10.1371/journal.ppat.0030104. PMC 1933454. PMID 17676996
- Shabman RS, Gulcicek EE, Stone KL, Basler CF (November 2011). "The Ebola virus VP24 protein prevents hnRNP C1/C2 binding to karyopherin α1 and partially alters its nuclear import". The Journal of Infectious Diseases. 204 Suppl 3: S904-10. doi:10.1093/infdis/jir323. PMC 3189985. PMID 21987768

Viral Diseases 6

- Viral Infection
- Viral Pneumonia
- Bronchiolitis
- HIV/AIDS

Viral diseases occur when an organism's body is attacked by pathogenic viruses, and virions. Some diseases are chicken pox, bronchiolitis, HIV/ AIDS, etc. All the aspects related to viral infections and diseases have been carefully analyzed in this chapter.

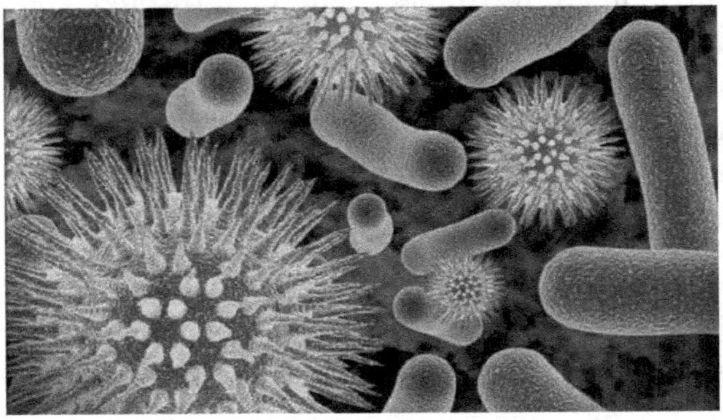

A viral disease is any illness or health condition caused by a virus.

Not all viral diseases are contagious. This means they aren't always spread from person to person. But many of them are. Common examples of contagious viral diseases include the flu, the common cold, HIV, and herpes. Other types of viral diseases spread through other means, such as the bite of an infected insect.

Respiratory Viral Diseases

Respiratory viral diseases are contagious and commonly affect the upper or lower parts of your respiratory tract.

Common symptoms of a respiratory viral disease include:

- Runny or stuffy nose.

- Coughing or sneezing.
- Fever.
- Body aches.

Examples:

Examples of respiratory diseases include:

- Flu.
- Common cold.
- Respiratory syncytial virus infection.
- Adenovirus infection.
- Parainfluenza virus infection.
- Severe acute respiratory syndrome (sars).

Transmission

Respiratory viruses are spread by droplets generated through coughing or sneezing. If someone with a viral illness coughs or sneezes nearby and you inhale these droplets, you may develop the disease. These viruses can also be spread through contaminated objects, such as doorknobs, tabletops, and personal items. If you touch one of these objects and then touch your nose or eyes, you could develop a disease.

Prevention

The best way to avoid respiratory viral diseases is to practice good personal hygiene. Wash your hands often, cover your mouth when you cough or sneeze, and limit your interactions with people who show symptoms of a respiratory condition.

There's also a vaccine that can help to reduce your risk of getting the seasonal flu.

Gastrointestinal Viral Diseases

Gastrointestinal viral diseases affect your digestive tract. The viruses that cause them are contagious and usually lead to a condition called gastroenteritis, also called the stomach flu.

Common symptoms of gastrointestinal viral diseases include:

- Abdominal cramps.
- Diarrhea.
- Vomiting.

Examples:

Examples of gastrointestinal viral diseases include:

- Norovirus infection.
- Rotavirus infection.
- Some adenovirus infections.
- Astrovirus infection.

Transmission

Gastrointestinal viruses are shed in the stool during bowel movements. Food or water that's been contaminated by feces can spread the virus to others. You can also get the virus from sharing utensils or personal objects with someone who has a virus.

Prevention

You can prevent gastrointestinal viral diseases by washing your hands often, especially after using the bathroom. Wiping down contaminated surfaces and not sharing personal items or eating utensils can also help.

There's also a vaccine for rotavirus that's recommended as part of a child's vaccination schedule.

Exanthematous Viral Disease

Exanthematous viruses cause skin rashes. Many of them cause additional symptoms as well. Many of the viruses in this category, such as the measles virus, are highly contagious.

Examples:

Examples of exanthematous viral diseases include:

- Measles.
- Rubella.
- Chickenpox/shingles.
- Roseola.
- Smallpox.
- Fifth disease.
- Chikungunya virus infection.

Transmission

- Many exanthematous viruses are spread through respiratory droplets from the cough or sneeze of someone with the virus.
- Other exanthematous viral diseases, such as chickenpox and smallpox, can be transmitted by coming into contact with fluid in broken skin lesions.
- Shingles only occurs in people who've had chickenpox at some point. It's a reactivation of the varicella-zoster virus that's been lying dormant in your cells.
- Chikungunya virus is spread through a mosquito bite and cannot be transmitted from person to person.

Prevention

Measles, rubella, chickenpox, shingles, and smallpox can all be prevented through vaccination. You can reduce your risk of a chikungunya virus infection by protecting yourself from mosquito bites.

Hepatic Viral Diseases

The hepatic viral diseases cause inflammation of the liver, known as viral hepatitis. The most common types of viral hepatitis are hepatitis A, B, and C.

It is worth noting that diseases caused by other viruses, such as cytomegalovirus and the yellow fever virus, can also affect the liver.

Examples:

Examples of hepatic viral diseases include:

- Hepatitis A.
- Hepatitis B.
- Hepatitis C.
- Hepatitis D.
- Hepatitis E.

Transmission

Many neurologic viruses are spread through the bite of an infected animal or bug, such as a mosquito or tick.

Other viruses, such poliovirus and other enteroviruses, are quite contagious and spread through close contact with someone with the virus. Contaminated objects can also contribute to the spread of these viruses.

Prevention

There's a vaccine for both poliovirus and the mumps virus, which can cause meningitis and encephalitis. Practicing good hygiene, avoiding close contact with those who have the virus, and protecting against insect bites can all help to reduce the spread of encephalitis and meningitis.

To reduce the risk of spreading rabies, keep your pets vaccinated and avoid approaching wild animals.

Cutaneous Viral Diseases

Cutaneous viral diseases cause lesions or papules to form on the skin. In many cases, these lesions can stick around for a long time or come back after disappearing for a while.

Examples:

Examples of cutaneous viral diseases include:

- Warts, including genital warts.
- Oral herpes.
- Genital herpes.
- Molluscum contagiosum.

Hemorrhagic Viral Diseases

Hemorrhagic viral diseases are severe conditions that involve damage to your circulatory system.

Symptoms of a hemorrhagic viral disease include:

- High fever.
- Body aches.
- Weakness.
- Bleeding under the skin.
- Bleeding from the mouth or ears.
- Bleeding in internal organs.

Examples:

Examples of viral hemorrhagic diseases include:

- Ebola.
- Lassa fever.
- Dengue fever.
- Yellow fever.
- Marburg hemorrhagic fever.
- Crimean-Congo hemorrhagic fever.

Transmission

Some hemorrhagic viral diseases, such as dengue fever and yellow fever, are spread through the bite of an infected insect.

Others, such as Ebola, are spread to other people through contact with the blood or other bodily fluid of someone with the virus. Lassa fever is spread through inhaling or consuming the dried feces or urine of a rodent with the virus.

Prevention

Researchers are in the process of developing vaccines for several hemorrhagic viruses. A yellow fever vaccine is currently available for people traveling to areas where yellow fever is common.

If you live or work in an area where viral hemorrhagic diseases are common, you can do the following to reduce your risk:

- Use proper protection, such as gloves, glasses, or a face shield, when working around people who have a virus.
- Avoid being bitten by insects, especially mosquitos and ticks, by wearing protective clothing or using insect repellent.
- Protect against rodent infestation by keeping food covered, removing garbage often, and making sure windows and doors are secured properly.

Neurologic Viral Diseases

Some viruses can infect the brain and surrounding tissues, causing neurologic viral diseases. This can result in a range of symptoms, including:

- Fever.

- Confusion.
- Drowsiness.
- Seizures.
- Coordination problems.

Examples:

Examples of neurologic viral diseases include:

- Polio.
- Viral meningitis.
- Viral encephalitis.
- Rabies.

Viral Infection

Viruses are small particles of genetic material (either DNA or RNA) that are surrounded by a protein coat. Some viruses also have a fatty "envelope" covering. They are incapable of reproducing on their own. Viruses depend on the organisms they infect (hosts) for their very survival. Viruses get a bad rap, but they also perform many important functions for humans, plants, animals, and the environment. For example, some viruses protect the host against other infections. Viruses also participate in the process of evolution by transferring genes among different species. In biomedical research, scientists use viruses to insert new genes into cells.

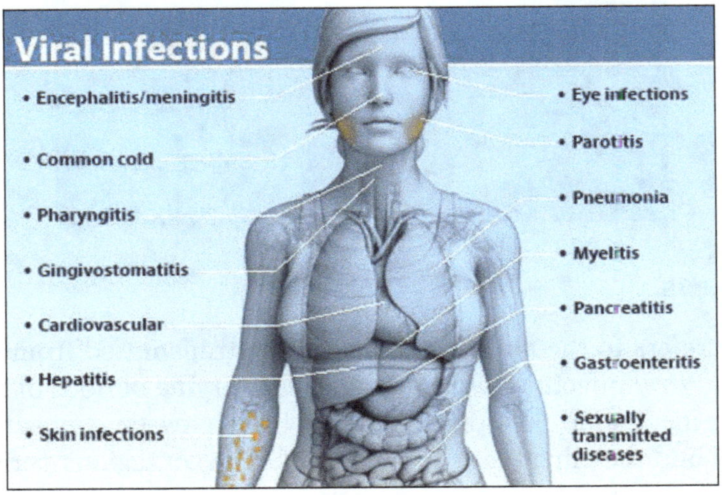

When most people hear the word "virus," they think of disease-causing (pathogenic) viruses such as the common cold, influenza, chickenpox, human immunodeficiency virus (HIV), and others. Viruses can affect many areas in the body, including the reproductive, respiratory, and gastrointestinal systems. They can also affect the liver, brain, and skin. Research reveals that that viruses are implicated in many cancers as well.

A viral infection is a proliferation of a harmful virus inside the body. Viruses cannot reproduce without the assistance of a host. Viruses infect a host by introducing their genetic material into the cells and hijacking the cell's internal machinery to make more virus particles. With an active viral infection, a virus makes copies of itself and bursts the host cell (killing it) to set the newly-formed virus particles free. In other cases, virus particles "bud" off the host cell over a period of time before killing the host cell. Either way, new virus particles are then free to infect other cells. Symptoms of the viral illness occur as a result of cell damage, tissue destruction, and the associated immune response.

Certain viruses - like the ones that cause chickenpox and cold sores - may be inactive or "latent" after the initial infection. For example, you may have a cold sore that erupts and then heals. The cold sore virus remains in your cells in a dormant state. At a later date, a trigger, such as stress, sunlight, or something else, may reactivate the virus and lead to new symptoms. The virus makes more copies of itself, releases new virus particles, and kills more host cells.

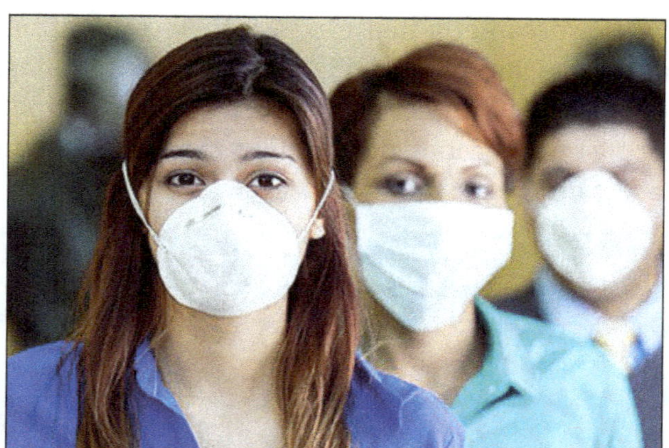

Contagiousness

Contagiousness refers to the ability of a virus to be transmitted from one person (or host) to another. Viral infections are contagious for varying periods of time depending on the virus. An incubation period refers to the time between exposure to a virus (or other pathogen) and the emergence of symptoms. The contagious period of a virus is not necessarily the same as the incubation period.

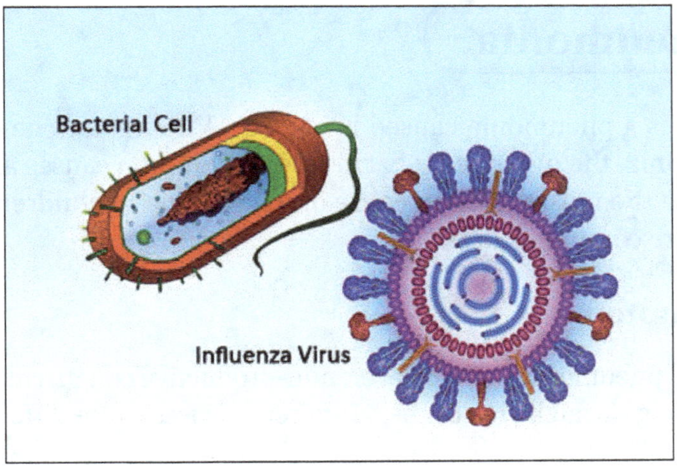

Bacteria vs. Virus

Viruses and bacteria are two types of potentially disease-causing (pathogenic) particles. Viruses are much smaller than bacteria and can't reproduce without the assistance of a host. Bacteria are capable of reproducing on their own. The symptoms of viral and bacterial illnesses are sometimes similar. A doctor can determine the underlying cause of an illness based on the patient's symptoms and other factors. Lab tests may help clarify whether an illness is due to a virus, bacteria, or other infectious agent or disease process.

Virus Transmission

Viruses can be transmitted in a variety of ways. Some viruses can spread through touch, saliva, or even the air. Other viruses can be transmitted through sexual contact or by sharing contaminated needles. Insects including ticks and mosquitoes can act as "vectors," transmitting a virus from one host to another. Contaminated food and water are other potential sources of viral infection.

Viral Pneumonia

Viral pneumonia is a pneumonia caused by a virus. Viruses are one of the two major causes of pneumonia, the other being bacteria; less common causes are fungi and parasites. Viruses are the most common cause of pneumonia in children, while in adults bacteria are a more common cause.

Signs and Symptoms

Symptoms of viral pneumonia include fever, non-productive cough, runny nose, and systemic symptoms (e.g. myalgia, headache). Different viruses cause different symptoms.

Causes

Common causes of viral pneumonia are:

- Influenza virus A and B.
- Respiratory syncytial virus (RSV).
- Human parainfluenza viruses (in children).

Rarer viruses that commonly result in pneumonia include:

- Adenoviruses (in military recruits).
- Metapneumovirus.
- Severe acute respiratory syndrome virus (SARS coronavirus).
- Middle East respiratory syndrome virus (MERS coronavirus).
- Hantaviruses.
- Novel coronavirus (2019-nCov).

Viruses that primarily cause other diseases, but sometimes cause pneumonia include:

- Herpes simplex virus (HSV), mainly in newborns or young children.
- Varicella-zoster virus (VZV).
- Measles virus.
- Rubella virus.
- Cytomegalovirus (CMV), mainly in people with immune system problems.
- Smallpox virus.
- Dengue virus.

The most commonly identified agents in children are respiratory syncytial virus, rhinovirus, human metapneumovirus, human bocavirus, and parainfluenza viruses.

Pathophysiology

Viruses must invade cells in order to reproduce. Typically, a virus will reach the lungs by traveling in droplets through the mouth and nose with inhalation. There, the virus invades the cells lining the airways and the alveoli. This invasion often leads to cell death either through direct killing by the virus or by self-destruction through apoptosis.

Further damage to the lungs occurs when the immune system responds to the infection. White blood cells, in particular lymphocytes, are responsible for activating a variety of chemicals (cytokines) which cause leaking of fluid into the alveoli. The combination of cellular destruction and fluid-filled alveoli interrupts the transportation of oxygen into the bloodstream.

In addition to the effects on the lungs, many viruses affect other organs and can lead to illness affecting many different bodily functions. Some viruses also make the body more susceptible to bacterial infection; for this reason, bacterial pneumonia often complicates viral pneumonia.

Prevention

The best prevention against viral pneumonia is vaccination against influenza, adenovirus, chickenpox, herpes zoster, measles, and rubella.

Treatment

In cases of viral pneumonia where influenza A or B are thought to be causative agents, patients who are seen within 48 hours of symptom onset may benefit from treatment with oseltamivir or zanamivir. Respiratory syncytial virus (RSV) has no direct acting treatments, but ribavirin in indicated for severe cases. Herpes simplex virus and varicella-zoster virus infections are usually treated with aciclovir, whilst ganciclovir is used to treat cytomegalovirus. There is no known efficacious treatment for pneumonia caused by SARS coronavirus, MERS coronavirus, adenovirus, hantavirus, or parainfluenza. Care is largely supportive.

Bronchiolitis

Bronchiolitis is blockage of the small airways in the lungs due to a viral infection. It usually only occurs in children less than two years of age. Symptoms may include fever, cough, runny nose, wheezing, and breathing problems. More severe cases may be associated with nasal flaring, grunting, or the skin between the ribs pulling in with breathing. If the child has not been able to feed properly, signs of dehydration may be present.

Bronchiolitis is usually the result of infection by respiratory syncytial virus (72% of cases) or human rhinovirus (26% of cases). Diagnosis is generally based on symptoms. Tests such as a chest X-ray or viral testing are not routinely needed.

There is no specific treatment. Supportive care at home is generally sufficient. Occasionally hospital admission for oxygen, support with feeding, or intravenous fluids is required. Tentative evidence supports nebulized hypertonic saline. Evidence for antibiotics, antivirals, bronchodilators, or nebulized epinephrine is either unclear or not supportive.

About 10% to 30% of children under the age of two years are affected by bronchiolitis at some point in time. It more commonly occurs in the winter in the Northern hemisphere. It is the leading cause of hospitalizations in those less than one year of age in the United States. The risk of death among those who are admitted to hospital is about 1%. Outbreaks of the condition were first described in the 1940s.

Signs and Symptoms

Bronchiolitis typically presents in children under two years old and is characterized by a constellation of respiratory symptoms that consists of fever, rhinorrhea, cough, wheeze, tachypnea and increased work of breathing such as nasal flaring or grunting that develops over one to three days. Crackles or wheeze are typical findings on listening to the chest with a stethoscope. The child may also experience apnea, or brief pauses in breathing. After the acute illness, it is common for the airways to remain sensitive for several weeks, leading to recurrent cough and wheeze.

Some signs of severe disease include:

- Poor feeding (less than half of usual fluid intake in preceding 24 hours).
- Significantly decreased activity.
- History of stopping breathing.
- Respiratory rate >70/min.
- Presence of nasal flaring and/or grunting.
- Severe chest wall recession (hoover's sign).
- Bluish skin.

Causes

The term usually refers to acute viral bronchiolitis, a common disease in infancy. This is most commonly caused by respiratory syncytial virus (RSV, also known as human pneumovirus). Other agents that cause this illness include human metapneumovirus, influenza, parainfluenza, coronavirus, adenovirus, rhinovirus and mycoplasma.

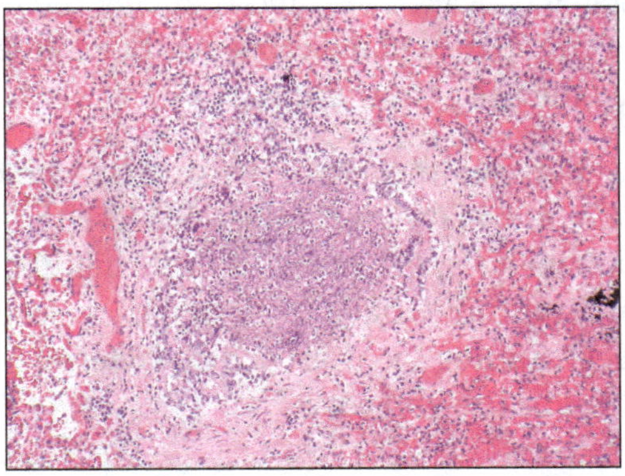

Acute inflammatory exudate occluding the lumen of the bronchiole and acute inflammation of part of the wall of the bronchiole.

Risk Factors

Children are at an increased risk for progression to severe respiratory disease if they have any of the following additional factors:

- Preterm infant (gestational age less than 37 weeks).
- Younger age at onset of illness (less than 3 months of age).
- Congenital heart disease.
- Immunodeficiency.
- Chronic lung disease.
- Neurological disorders.
- Tobacco smoke exposure.

Diagnosis

The diagnosis is typically made by clinical examination. Chest X-ray is sometimes useful to exclude bacterial pneumonia, but not indicated in routine cases. Chest x-ray may also be useful in people with impending respiratory failure. Additional testing such as blood cultures, complete blood count, and electrolyte analyses are not recommended for routine use although may be useful in children with multiple comorbidities or signs of sepsis or pneumonia.

Testing for the specific viral cause can be done but has little effect on management and thus is not routinely recommended. RSV testing by direct immunofluorescence testing on nasopharyngeal aspirate had a sensitivity of 61% and specificity of 89%.

Identification of those who are RSV-positive can help for disease surveillance, grouping ("cohorting") people together in hospital wards to prevent cross infection, predicting whether the disease course has peaked yet, and reducing the need for other diagnostic procedures (by providing confidence that a cause has been identified). Identification of the virus may help reduce the use of antibiotics.

Infants with bronchiolitis between the age of two and three months have a second infection by bacteria (usually a urinary tract infection) less than 6% of the time. When further evaluated with a urinalysis, infants with bronchiolitis had a concomitant UTI 0.8% of the time. Preliminary studies have suggested that elevated procalcitonin levels may assist clinicians in determining the presence of bacterial co-infection, which could prevent unnecessary antibiotic use and costs.

Differential Diagnosis

There are many childhood illnesses that can present with respiratory symptoms, particularly persistent cough and wheezing. Bronchiolitis may be differentiated from some of these by the characteristic pattern of preceding febrile upper respiratory tract symptoms lasting for 1 to 3 days followed by the persistent cough, tachypnea, and wheezing. However, some infants may present without fever (30% of cases) or may present with apnea without other signs or with poor weight gain prior to onset of symptoms. In such cases, additional laboratory testing and radiographic imaging may be useful. The following are some other diagnoses to consider in an infant presenting with signs of bronchiolitis:

- Asthma and reactive airway disease.
- Bacterial pneumonia.
- Congenital heart disease.
- Heart failure.
- Whooping cough.
- Allergic reaction.
- Cystic fibrosis.
- Chronic pulmonary disease.
- Foreign body aspiration.
- Vascular ring.

Prevention

Prevention of bronchiolitis relies strongly on measures to reduce the spread of the viruses that cause respiratory infections (that is, handwashing, and avoiding exposure to those symptomatic with respiratory infections). Guidelines are mixed on the use of gloves, aprons, or personal protective equipment. In addition to good hygiene, an improved immune system is a great tool for prevention.

One way to improve the immune system is to feed the infant with breast milk, especially during the first month of life. Respiratory infections were shown to be significantly less common among breastfed infants and fully breastfed RSV-positive hospitalized infants had shorter hospital stays than non or partially breastfed infants. Guidelines recommend exclusive breastfeeding for infants for the first 6 months of life.

Palivizumab, a monoclonal antibody against RSV, can be administered to prevent bronchiolitis to infants less than one year of age that were born very prematurely or that have underlying heart disease or chronic lung disease of prematurity. Passive immunization therapy requires monthly injections during winter. Otherwise healthy premature infants that were born after a gestational age of 29 weeks should not be administered palivizumab as the harms outweigh the benefits. Passive protection through the administration of other novel monoclonal antibodies is also under evaluation.

The development of immunizations for RSV are being developed but there are none available currently.

Tobacco smoke exposure has been shown to increase both the rates of lower respiratory disease in infants as well as the risk and severity of bronchiolitis. Tobacco smoke lingers in the environment for prolonged periods and on clothing even when smoking outside the home. Guidelines recommend that parents be fully educated on the risks of tobacco smoke exposure on children with bronchiolitis.

Management

Treatment of bronchiolitis is usually focused on the hydration and symptoms instead of the infection itself since the infection will run its course and complications are typically from the symptoms themselves. Without active treatment, half of cases will go away in 13 days and 90% in three weeks. Children with severe symptoms, especially poor feeding or dehydration, may be considered for hospital admission. Oxygen saturation under 90%-92% as measured with pulse oximetry is also frequently used as an indicator of need for hospitalization. High-risk infants, apnea, cyanosis, malnutrition, and diagnostic uncertainty are additional indications for hospitalization.

Most guidelines recommend sufficient fluids and nutritional support for affected children. Measures for which the recommendations were mixed include nebulized hypertonic saline, nebulized epinephrine, and nasal suctioning. Treatments which the evidence does not support include salbutamol, steroids, antibiotics, antivirals, heliox, continuous positive airway pressure (CPAP), chest physiotherapy, and cool mist or steam inhalation.

Diet

Maintaining hydration is an important part of management of bronchiolitis. Infants with mild pulmonary symptoms may require only observation if feeding is unaffected. However, oral intake may be affected by nasal secretions and increased work of breathing. Poor feeding or dehydration, defined as less than 50% of usual intake, is often cited as an indication for hospital admission. Guidelines recommend the use of nasogastric or intravenous fluids in children with bronchiolitis who cannot maintain usual oral intake. The risk of health care caused hyponatremia and fluid retention are minimal with the use of isotonic fluids such as normal saline, breast milk, or formula.

Oxygen

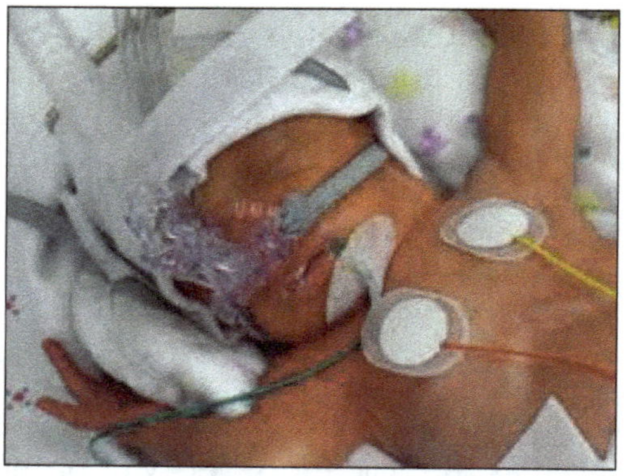

A newborn wearing a nasal CPAP device.

Inadequate oxygen supply to the tissue is one of the main concerns during severe bronchiolitis and oxygen saturation is often closely associated with both the need for hospitalization and continued length of hospital stay in children with bronchiolitis. However, oxygen saturation is a poor predictor of respiratory distress. Accuracy of pulse oximetry is limited in the 76% to 90% range and there is weak correlation between oxygen saturation and respiratory distress as brief hypoxemia is common in healthy infants. Additionally, pulse oximetry is associated with frequent false alarms and parental stress and fatigue. Clinicians may choose not to given additional oxygen to children with bronchiolitis if their oxygen saturation is above 90%. Additionally, clinicians may choose not to use continuous pulse oximetry in these people.

When choosing to use oxygen therapy for a child with bronchiolitis, there is evidence that home oxygen may reduce hospitalization rate and length of stay although readmission rates and follow-up visits are increased. Also, the use of humidified, heated, high-flow nasal cannula may be a safe initial therapy to decrease work of breathing and need for intubation. However, evidence is lacking regarding the use of high-flow nasal cannula compared to standard oxygen therapy or continuous positive airway pressure. These practices may still be used in severe cases prior to intubation.

Blood gas testing is not recommended for people hospitalized with the disease and is not useful in the routine management of bronchiolitis. People with severe worsening respiratory distress or impending respiratory failure may be considered for capillary blood gas testing.

Hypertonic Saline

Guidelines recommend against the use of nebulized hypertonic saline in the emergency department for children with bronchiolitis but it may be give to children who are hospitalized.

Nebulized hypertonic saline (3%) has limited evidence of benefit and previous studies lack consistency and standardization. A 2017 review found tentative evidence that it reduces the risk of hospitalization, duration of hospital stay, and improved the severity of symptoms. The majority of evidence suggests that hypertonic saline is safe and effective at improving respiratory symptoms of mild to moderate bronchiolitis after 24 hours of use. However, it does not appear effective in reducing the rate of hospitalization when used in the emergency room or other outpatient settings in which length of therapy is brief. Side effects were mild and resolved spontaneously.

Bronchodilators

Guidelines recommend against the use of bronchodilators in children with bronchiolitis as evidence does not support a change in outcomes with such use.

Several studies have shown that bronchodilation with β-adrenergic agents such as salbutamol may improve symptoms briefly but do not affect the overall course of the illness or reduce the need for hospitalization. However, there are conflicting recommendations about the use of a trial of a bronchodilator, especially in those with history of previous wheezing, due to the difficulty with assessing an objective improvement in symptoms. Additionally, there are adverse effects to the use of bronchodilators in children such as tachycardia and tremors as well as adding increased financial expenses.

Anticholinergic inhalers, such as ipratropium bromide, have a modest short term effect at best and are not recommended for treatment.

Epinephrine

The current state of evidence suggests that nebulized epinephrine is not indicated for children with bronchiolitis except as a trial of rescue therapy for severe cases.

Epinephrine is an α and β adrenergic agonist that has been used to treat other upper respiratory tract illnesses, such as croup, as a nebulized solution. A Cochrane meta-analysis in 2011 found no benefit to the use of epinephrine in the inpatient setting and suggested that there may be utility in the outpatient setting in reducing the rate of hospitalization. However, current guidelines do not support the outpatient use of epinephrine given the lack of substantial sustained benefit. A 2017 review found inhaled epinephrine with corticosteroids did not change the need for hospitalization or the time spent in hospital. Other studies suggest a synergistic effect of epinephrine with corticosteroids but have not consistently demonstrated benefits in clinical trials. Guidelines recommend against its use currently.

Unclear Evidence

Currently other medications do not yet have evidence to support their use, although they have been studied for use in bronchiolitis. Experimental trials with novel antiviral medications in adults are promising but it remains unclear if the same benefit will be present.

- Surfactant had favorable effects for severely critical infants on duration of mechanical ventilation and ICU stay however studies were few and small.
- Chest physiotherapy, such as vibration or percussion, to promote airway clearance may slightly reduce duration of oxygen therapy but there is a lack of evidence that demonstrates any other benefits. People with difficulty clearing secretions due to underlying disorders such as spinal muscle atrophy or severe tracheomalacia may be considered for chest physiotherapy.
- Suctioning of the nares may provide temporarily relief of nasal congestion but deep suctioning of the nasopharynx has been shown prolong length of hospital stay in infants. Upper airway suctioning may be considered in people with respiratory distress, feeding difficulties, or infants presenting with apnea.
- Heliox, a mixture of oxygen and the inert gas helium, may be beneficial in infants with severe acute RSV bronchiolitis who require CPAP but overall evidence is lacking.
- DNAse has not been found to be effective but might play a role in severe bronchiolitis complicated by atelectasis.
- There are no systematic reviews or controlled trials on the effectiveness of nasal decongestants, such as xylometazoline, for the treatment of bronchiolitits.
- Overall evidence is insufficient to support the use of alternative medicine. There is tentative evidence for Chinese herbal medicine, vitamin D, N-acetylcysteine, and magnesium but this is insufficient to recommend their use.

Non-effective Treatments

- Ribavirin is an antiviral drug which does not appear to be effective for bronchiolitis.
- Antibiotics are often given in case of a bacterial infection complicating

bronchiolitis, but have no effect on the underlying viral infection and their benefit is not clear. The risks of bronchiolitis with a concomitant serious bacterial infection among hospitalized febrile infants is minimal and work-up and antibiotics are not justified. Azithromycin adjuvant therapy may reduce the duration of wheezing and coughing in children with bronchiolitis but has not effect on length of hospital stay or duration of oxygen therapy.

- Corticosteroids, although useful in other respiratory disease such as asthma and croup, have no proven benefit in bronchiolitis treatment and are not advised. Additionally, corticosteroid therapy in children with bronchiolitis may prolong viral shedding and transmissibility. The overall safety of corticosteroids is questionable.
- Leukotriene inhibitors, such as montelukast, have not been found to be beneficial and may increase adverse effects.
- Immunoglobulins are of unclear benefit.

HIV/AIDS

Human immunodeficiency virus infection and acquired immune deficiency syndrome (HIV/AIDS) is a spectrum of conditions caused by infection with the human immunodeficiency virus (HIV). Following initial infection a person may not notice any symptoms, or may experience a brief period of influenza-like illness. Typically, this is followed by a prolonged period with no symptoms. If the infection progresses, it interferes more with the immune system, increasing the risk of developing common infections such as tuberculosis, as well as other opportunistic infections, and tumors which are otherwise rare in people who have normal immune function. These late symptoms of infection are referred to as acquired immunodeficiency syndrome (AIDS). This stage is often also associated with unintended weight loss.

HIV is spread primarily by unprotected sex (including anal and oral sex), contaminated blood transfusions, hypodermic needles, and from mother to child during pregnancy, delivery, or breastfeeding. Some bodily fluids, such as saliva, sweat and tears, do not transmit the virus. HIV is a member of the group of viruses known as retroviruses. Methods of prevention include safe sex, needle exchange programs, treating those who are infected, pre- and post-exposure prophylaxis, and male circumcision. Disease in a baby can often be prevented by giving both the mother and child antiretroviral medication. There is no cure or vaccine; however, antiretroviral treatment can slow the course of the disease and may lead to a near-normal life expectancy. Treatment is recommended as soon as the diagnosis is made. Without treatment, the average survival time after infection is 11 years.

In 2018 about 37.9 million people were living with HIV and it resulted in 770,000 deaths. An estimated 20.6 million of these live in eastern and southern Africa. Between the time that AIDS was identified (in the early 1980s) and 2018, the disease caused an estimated 32 million deaths worldwide. HIV/AIDS is considered a pandemic—a disease outbreak which is present over a large area and is actively spreading. HIV originated in west-central Africa during the late 19th or early 20th century. AIDS was first recognized by the United States Centers for Disease Control and Prevention (CDC) in 1981 and its cause—HIV infection—was identified in the early part of the decade.

HIV/AIDS has had a large impact on society, both as an illness and as a source of discrimination. The disease also has large economic impacts. There are many misconceptions about HIV/AIDS, such as the belief that it can be transmitted by casual non-sexual contact. The disease has become subject to many controversies involving religion, including the Catholic Church's position not to support condom use as prevention. It has attracted international medical and political attention as well as large-scale funding since it was identified in the 1980s.

Signs and Symptoms

There are three main stages of HIV infection: acute infection, clinical latency, and AIDS.

Acute Infection

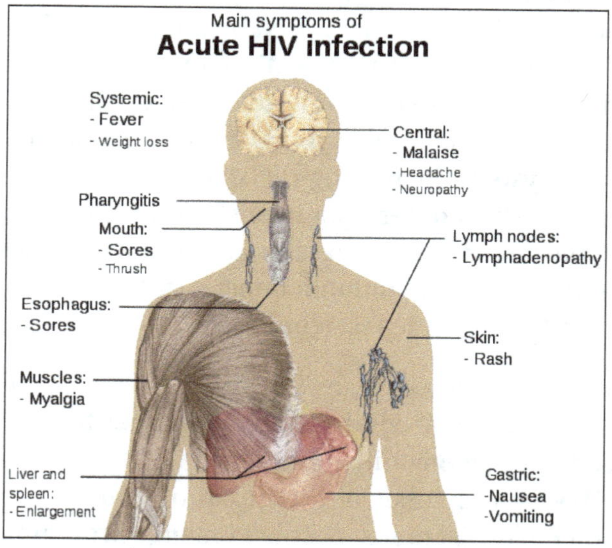

Main symptoms of acute HIV infection.

The initial period following the contraction of HIV is called acute HIV, primary HIV or acute retroviral syndrome. Many individuals develop an influenza-like illness or a mononucleosis-like illness 2–4 weeks after exposure while others have no significant symptoms. Symptoms occur in 40–90% of cases and most commonly include fever, large tender lymph nodes, throat inflammation, a rash, headache, tiredness, and/or sores of the mouth and genitals. The rash, which occurs in 20–50% of cases, presents itself on the trunk and is maculopapular, classically. Some people also develop opportunistic infections at this stage. Gastrointestinal symptoms, such as vomiting or diarrhea may occur. Neurological symptoms of peripheral neuropathy or Guillain–Barré syndrome also occurs. The duration of the symptoms varies, but is usually one or two weeks.

Owing to their nonspecific character, these symptoms are not often recognized as signs of HIV infection. Even cases that do get seen by a family doctor or a hospital are often misdiagnosed as one of the many common infectious diseases with overlapping symptoms. Thus, it is recommended that HIV be considered in people presenting with an unexplained fever who may have risk factors for the infection.

Clinical Latency

The initial symptoms are followed by a stage called clinical latency, asymptomatic HIV, or chronic HIV. Without treatment, this second stage of the natural history of HIV infection can last from about three years to over 20 years (on average, about eight years). While typically there are few or no symptoms at first, near the end of this stage many people experience fever, weight loss, gastrointestinal problems and muscle pains. Between 50% and 70% of people also develop persistent generalized lymphadenopathy, characterized by unexplained, non-painful enlargement of more than one group of lymph nodes (other than in the groin) for over three to six months.

Although most HIV-1 infected individuals have a detectable viral load and in the absence of treatment will eventually progress to AIDS, a small proportion (about 5%) retain high levels of $CD4^+$ T cells (T helper cells) without antiretroviral therapy for more than five years. These individuals are classified as "HIV controllers" or long-term non-progressors (LTNP). Another group consists of those who maintain a low or undetectable viral load without anti-retroviral treatment, known as "elite controllers" or "elite suppressors". They represent approximately 1 in 300 infected persons.

Acquired Immunodeficiency Syndrome

Acquired immunodeficiency syndrome (AIDS) is defined in terms of either a $CD4^+$ T cell count below 200 cells per μL or the occurrence of specific diseases in association with an HIV infection. In the absence of specific treatment, around half of people infected with HIV develop AIDS within ten years. The most common initial conditions that alert to the presence of AIDS are pneumocystis pneumonia (40%), cachexia in the form of HIV wasting syndrome (20%), and esophageal candidiasis. Other common signs include recurrent respiratory tract infections.

Opportunistic infections may be caused by bacteria, viruses, fungi, and parasites that are normally controlled by the immune system. Which infections occur depends partly on what organisms are common in the person's environment. These infections may affect nearly every organ system.

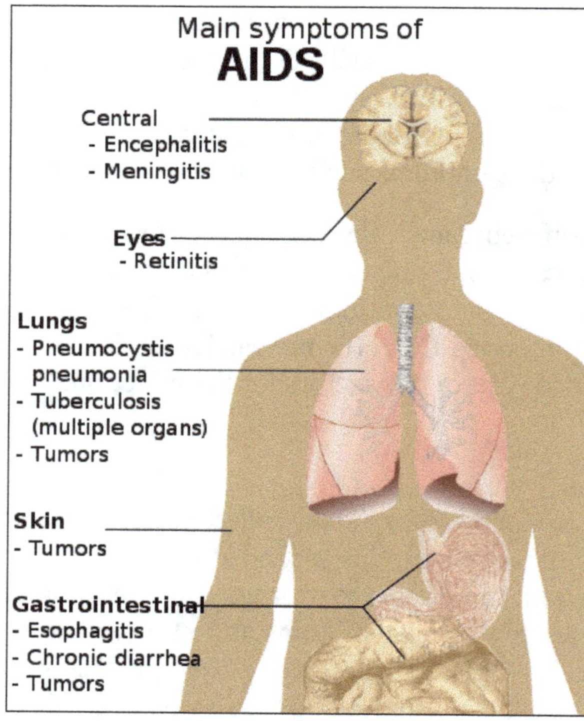

Main symptoms of AIDS.

People with AIDS have an increased risk of developing various viral-induced cancers, including Kaposi's sarcoma, Burkitt's lymphoma, primary central nervous system lymphoma, and cervical cancer. Kaposi's sarcoma is the most common cancer, occurring in 10% to 20% of people with HIV. The second-most common cancer is lymphoma, which is the cause of death of nearly 16% of people with AIDS and is the initial sign of AIDS in 3% to 4%. Both these cancers are associated with human herpesvirus 8 (HHV-8). Cervical cancer occurs more frequently in those with AIDS because of its association with human papillomavirus (HPV). Conjunctival cancer (of the layer that lines the inner part of eyelids and the white part of the eye) is also more common in those with HIV.

Additionally, people with AIDS frequently have systemic symptoms such as prolonged fevers, sweats (particularly at night), swollen lymph nodes, chills, weakness, and unintended weight loss. Diarrhea is another common symptom, present in about 90% of people with AIDS. They can also be affected by diverse psychiatric and neurological symptoms independent of opportunistic infections and cancers.

Transmission

Table: Average per act risk of getting HIV by exposure route to an infected source.

Exposure route	Chance of infection
Blood transfusion	90%
Childbirth (to child)	25%

Needle-sharing injection drug use	0.67%
Percutaneous needle stick	0.30%
Receptive anal intercourse	0.04–3.0%
Insertive anal intercourse	0.03%
Receptive penile-vaginal intercourse	0.05–0.30%
Insertive penile-vaginal intercourse	0.01–0.38%
Receptive oral intercourse	0–0.04%
Insertive oral intercourse	0–0.005%

HIV is spread by three main routes: sexual contact, significant exposure to infected body fluids or tissues, and from mother to child during pregnancy, delivery, or breast-feeding (known as vertical transmission). There is no risk of acquiring HIV if exposed to feces, nasal secretions, saliva, sputum, sweat, tears, urine, or vomit unless these are contaminated with blood. It is also possible to be co-infected by more than one strain of HIV—a condition known as HIV superinfection.

References

- Viral-diseases, health: healthline.com, Retrieved 18 May, 2020
- Viral-infections: onhealth.com, Retrieved 05 January, 2020
- Ruuskanen, O; Lahti, E; Jennings, LC; Murdoch, DR (2011-04-09). "Viral pneumonia". Lancet. 377 (9773): 1264–75. doi:10.1016/S0140-6736(10)61459-6. PMID 21435708
- Friedman JN, Rieder MJ, Walton JM, Canadian Paediatric Society, Acute Care Committee, Drug Therapy and Hazardous Substances (November 2014). "Bronchiolitis: Recommendations for diagnosis, monitoring and management of children one to 24 months of age". Paediatrics & Child Health. 19 (9): 485–98. doi:10.1093/pch/19.9.485. PMC 4235450. PMID 25414585
- Kallings LO (March 2008). "The first postmodern pandemic: 25 years of HIV/AIDS". Journal of Internal Medicine. 263 (3): 218–43. doi:10.1111/j.1365-2796.2007.01910.x. PMID 18205765
- Dennehy PH (2000). "Transmission of rotavirus and other enteric pathogens in the home". Pediatr. Infect. Dis. J. 19 (10 Suppl): S103–5. doi:10.1097/00006454-200010001-00003. PMID 11052397
- Viral-diseases-of-animals: iaszoology.com, Retrieved 19 February, 2020

PERMISSIONS

All chapters in this book are published with permission under the Creative Commons Attribution Share Alike License or equivalent. Every chapter published in this book has been scrutinized by our experts. Their significance has been extensively debated. The topics covered herein carry significant information for a comprehensive understanding. They may even be implemented as practical applications or may be referred to as a beginning point for further studies.

We would like to thank the editorial team for lending their expertise to make the book truly unique. They have played a crucial role in the development of this book. Without their invaluable contributions this book wouldn't have been possible. They have made vital efforts to compile up to date information on the varied aspects of this subject to make this book a valuable addition to the collection of many professionals and students.

This book was conceptualized with the vision of imparting up-to-date and integrated information in this field. To ensure the same, a matchless editorial board was set up. Every individual on the board went through rigorous rounds of assessment to prove their worth. After which they invested a large part of their time researching and compiling the most relevant data for our readers.

The editorial board has been involved in producing this book since its inception. They have spent rigorous hours researching and exploring the diverse topics which have resulted in the successful publishing of this book. They have passed on their knowledge of decades through this book. To expedite this challenging task, the publisher supported the team at every step. A small team of assistant editors was also appointed to further simplify the editing procedure and attain best results for the readers.

Apart from the editorial board, the designing team has also invested a significant amount of their time in understanding the subject and creating the most relevant covers. They scrutinized every image to scout for the most suitable representation of the subject and create an appropriate cover for the book.

The publishing team has been an ardent support to the editorial, designing and production team. Their endless efforts to recruit the best for this project, has resulted in the accomplishment of this book. They are a veteran in the field of academics and their pool of knowledge is as vast as their experience in printing. Their expertise and guidance has proved useful at every step. Their uncompromising quality standards have made this book an exceptional effort. Their encouragement from time to time has been an inspiration for everyone.

The publisher and the editorial board hope that this book will prove to be a valuable piece of knowledge for students, practitioners and scholars across the globe.

INDEX

www.ingramcontent.com/pod-product-compliance
Lightning Source LLC
LaVergne TN
LVHW080629210926
848485LV00003B/232
9781639894437